T0231108

BIFURCATIONS & INSTABILITIES IN GEOMECHANICS

PROCEEDINGS OF THE INTERNATIONAL WORKSHOP ON
BIFURCATIONS & INSTABILITIES IN GEOMECHANICS, IWBI 2002,
2-5 JUNE 2002, MINNEAPOLIS MN, USA

Bifurcations & Instabilities in Geomechanics

Edited by

J.F. Labuz & A. Drescher
University of Minnesota, Minneapolis MN, USA

A.A. BALKEMA PUBLISHERS LISSE / ABINGDON / EXTON (PA) / TOKYO

Cover: Schematic diagram of the University of Minnesota Plane Strain Apparatus; results from a plane-strain test on Berea sandstone (from the Geomechanics Laboratory, Department of Civil Engineering, University of Minnesota).

Published by: A.A. Balkema, a member of Swets & Zeitlinger Publishers
www.balkema.nl and www.szp.swets.nl

For the complete set of two volumes: ISBN 90 5809 563 0

Printed in the Netherlands

Table of Contents

4. Experimental studies

5. Non-local models and coupled effects

Preface

Several seminal papers on localization and related instabilities were published in the 1970's and early 1980's, and bifurcation theory was becoming a topical subject among a small group of engineers in the field of geomechanics. To assemble this modest community and provide a forum for discussion, the first International Workshop on Localization of Soils was held in Karlsruhe, Germany, February 1988. It was followed by the second international workshop in Gdansk, Poland, September 1989. The first two events emphasized fundamental aspects of bifurcation theory applied to soil mechanics. The third international workshop was held in Aussois, France, September 1993, and the topics were extended to include rock mechanics. The sites for the fourth and fifth workshops were Gifu, Japan (September 1997) and Perth, Australia (November 1999). The University of Minnesota was selected as the host for the sixth workshop, which was held on June 2-5, 2002 and named the International Workshop on Bifurcations & Instabilities in Geomechanics (IWBI 2002).

The common themes throughout the past workshops have been modeling of bifurcation, failure of geomaterials and structures, advanced analytical, numerical and experimental techniques, and application and development of generalized continuum models. The objective of IWBI 2002 was to continue these themes by bringing together international researchers and practitioners dealing with bifurcations and instabilities in geomechanics, specifically to collect and debate the developments and applications that have taken place the past few years. The scope of the workshop was broadened to include not only localization and other bifurcation modes, but also related instabilities, for example, due to softening (inherent or geometric) and loading conditions (static rather than kinematic), among others. This was reflected in the name "International Workshop on Bifurcations and Instabilities in Geomechanics." The venue for IWBI 2002 was the campus of St. John's University in Collegeville, Minnesota, about 100 km from the University of Minnesota. Surrounded by woodlands and lakes, St. John's provided a wonderful environment for discussions and interactions.

IWBI 2002 was attended by 74 participants representing 15 countries; 45 presentations were given over three days. The papers in this book are a sampling of the presentations. The development of the program was aided by the Scientific Committee consisting of F. Darve, H. Muhlhaus, F. Oka, I. Vardoulakis, and D. Muir Wood, with input from the Advisory Committee consisting of T. Adachi, J.P. Bardet, R. de Borst, J. Desrues, G. Gudehus, D. Kolymbas, R. Michalowski, Z. Mroz, R. Nova, S. Pietruszczak, and J. Rudnicki. Their efforts contributed immensely to the workshop.

IWBI 2002 was sponsored by the National Science Foundation, Civil & Mechanical Systems Division, Geomechanics and Geotechnical Systems. Additional support was provided by (in alphabetical order) BP America, Houston, TX; Itasca Consulting Group, Minneapolis, MN; MTS Systems, Eden Prairie, MN; Schlumberger Cambridge Research, Cambridge, UK; Schlumberger Engineering Applications, Sugar Land, TX; Shell International Exploration and Productions, Houston, TX; University of Minnesota, Institute of Technology and Department of Civil Engineering, Minneapolis, MN.

Joseph F. Labuz & Andrew Drescher

Organisation

The June 2-5, 2002 International Workshop on Bifurcations & Instabilities in Geomechanics was sponsored by the National Science Foundation, Civil & Mechanical Systems
Division, Geomechanics and Geotechnical Systems. Additional support was provided
by (in alphabetical order) BP America, Houston, TX; Itasca Consulting Group, Minneapolis, MN; MTS Systems, Eden Prairie, MN; Schlumberger Cambridge Research,
Cambridge, UK; Schlumberger Engineering Applications, Sugar Land, TX; Shell International Exploration and Productions, Houston, TX; University of Minnesota, Institute
of Technology and Department of Civil Engineering, Minneapolis, MN.

Scientific Committee: F. Darve (France), H. Muhlhaus (Australia), F. Oka (Japan), I.
Vardoulakis (Greece), D. Muir Wood (UK)

Advisory Committee: T. Adachi, (Japan), J.P. Bardet (USA), R. de Borst (The Neth-
erlands), J. Desrues (France), G. Gudehus (Germany), D. Kolymbas (Austria), R.
Michalowski (USA), Z. Mroz (Poland), R. Nova (Italy), S. Pietruszczak (Canada), J.
Rudnicki (USA)

1. Critical states and bifurcation conditions

Bifurcations & Instabilities in Geomechanics, – Labuz & Drescher (eds.)
© 2003 Swets & Zeitlinger, Lisse, ISBN 90 5809 563 0

The failure concept in Soil Mechanics revisited

R. Nova
Milan University of Technology (Politecnico)

ABSTRACT: When applied to porous materials characterised by a non- associate flow rule such as soils, the apparently self evident concept of failure becomes less neat. Although it is possible to define a convenient limit locus beyond which stress states are not feasible for a given material, unlimited strains may occur even for stress states within such locus, provided convenient loading conditions are imposed and controlling parameters appropriately chosen.

It is shown that such type of failures can occur in the hardening regime, provided the flow rule is non-associate and the stiffness matrix is not positive-definite. Failures due to loss of load control in undrained tests or shear or compaction band occurrence are examples.

The related concepts of hardening ad softening are discussed. It is shown that, contrary to primitive expectations, hardening may occur even with decreasing deviator stress, while softening may be associated to a load increment.

1 INTRODUCTION

In elementary Soil Mechanics books, the shear strength of geomaterials is defined in terms of the Coulomb-Mohr failure condition. In the principal stress space, the failure locus is the lateral surface of a pyramid of equation

$$\left|\sigma_i' - \sigma_j'\right| = 2c'\cos\phi' + (\sigma_i' + \sigma_j')\sin\phi' \tag{1}$$

where c' and ϕ' are the cohesion and the friction angle of the soil and σ_k' are the effective principal stresses.

This implies that the stress states for which the corresponding stress points lie on that surface are 'at failure', that means that unlimited strains can take place under constant stress. As a consequence, no stress point outside the locus can be reached. On the contrary, all stress states strictly within this locus are 'safe', which means that sufficiently small stress perturbations generate small strains only.

Unfortunately such an elementary concept of failure is contradicted by experimental evidence: 'failure', i.e. unlimited strains, can occur, under special conditions, even when the stress point is within the domain usually considered as 'safe'.

In order to show that, it is enough to consider a specimen of loose saturated sand in axisymmetric (so called 'triaxial') conditions. If a drained test at constant cell pressure is performed, the deviatoric stress, q ,increases with axial strain, ε_a , approaching a horizontal asymptote, as shown in Fig.1. The stress ratio η_f, defined as the ratio between q and the isotropic effective pressure, p', at this asymptotic level, is independent of the lateral confining pressure and corresponds to a friction angle of about 30°.

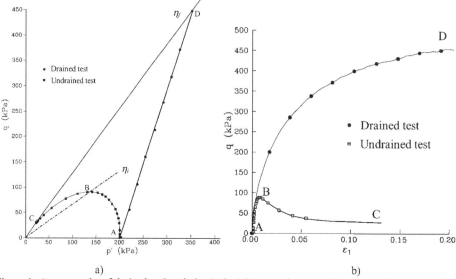

a)   b)

Figure 1. a) stress paths of drained and undrained triaxial tests on loose sand specimens b) corresponding stress-strain relationship

If, on the contrary, we perform an undrained test under displacement control and constant confining pressure, the pore water pressure increases with increasing strains, while the stress deviator reaches a peak and then dramatically decreases, eventually approaching a horizontal asymptote from above. The stress ratio corresponding to this final stage is again η_f, while the stress ratio at which the peak occurs, η_i, is much lower and corresponds to a mobilized friction angle of about 16°. Since strains at peak are still very small, the area of the right section of the specimen is almost the same as the initial one. Deviator stresses vary proportionally to the vertical load, therefore, and a peak in q corresponds to a peak in the axial load. It is clear then that, if an undrained test is performed on a theoretically identical specimen by monotonically increasing the vertical load, all the rest being unchanged, when the stress ratio reaches η_i, a loss of control of the specimen occurs and theoretically unlimited strains take place.

In geotechnical engineering, the peak in the undrained test is often considered to be a point of hardening-softening transition. This fact has a serious shortcoming, however. Softening is certainly associated to a loss of load control, as that shown by the experimental results, but also to a negative hardening modulus, if the terms hardening and softening are used as in the theory of plasticity. But this is not the case at the undrained peak. For an elastoplastic strainhardening material, the hardening modulus H is defined as follows:

$$d\varepsilon_{ij}^p = \frac{1}{H} \frac{\partial g}{\partial \sigma'_{ij}} \frac{\partial f}{\partial \sigma'_{hk}} d\sigma'_{hk} \tag{2}$$

$$H \equiv -\frac{\partial f}{\partial \psi_l} \frac{\partial \psi_l}{\partial \varepsilon_{rs}^p} \frac{\partial \varepsilon_{rs}^p}{\partial \sigma'_{rs}} \tag{3}$$

where g is the plastic potential, f is the loading function and ψ_l are the internal variables.

As it can be derived from Equation 3, the hardening modulus depends only on the state of stress and the previous plastic strain history of the material, but does not depend on the stress rate direction. The hardening modulus at the peak point is therefore the same in drained and undrained conditions. Since it is positive in the former case, it must be so also in the latter one.

Di Prisco et al. (1995) experimentally demonstrated this fact. Fig 2 shows the behavior of a loose sand specimen that is first loaded in undrained conditions up to the peak. Then, after pore water pressure equalization, drainage is allowed for and the axial load increased. As it is apparent, the behavior is typically hardening and no phenomenon of loss of control occurs. The response of the material depends therefore on the stress rate direction and can be associated either to an increase or a decrease of the stress deviator, depending on the kinematic control (volumetric strain allowed for or not). The hardening modulus must be positive in both cases, however, since it is independent of the stress rate direction and load control is possible in the drained case.

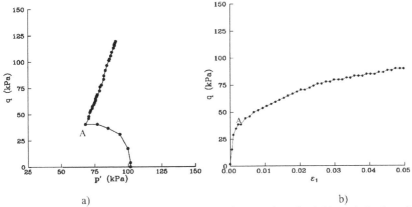

a)   b)

Figure 2. a) stress path and b) stress-strain relationship for a specimen loaded in undrained conditions up to peak and then allowed to drain for further axial loading- After di Prisco et al. (1995)

We can therefore conclude that, under special loading conditions and type of load control, failure of the specimen can occur at a stress point that is well within what is normally considered as a 'safe' domain. This is not associated to a hardening-softening transition, but it must be related to other properties that are hidden in the constitutive relationship of the material.

In the following, the conditions for such failures to occur will be recalled. It will be shown that failure can be associated to the concept of loss of control. Theoretically unlimited strain rates can take place, in fact, as a consequence of infinitesimal loading perturbations. Failure within the 'safe' domain is therefore associated to a homogeneous bifurcation, such as the unlimited pore water pressure generation in undrained tests. Even the occurrence of shear or compaction bands can be treated in this framework. This will eventually lead us to reconsidering the meaning of hardening and softening concepts, as traditionally perceived. It will be shown that, contrary to primitive expectations, hardening may occur even with decreasing deviator stress, while softening may be associated to a load increment.

2 CONDITIONS FOR LOSS OF CONTROL ('LATENT INSTABILITY')

The analysis that will be presented here (Nova (1989), (1994)) is based on the hypothesis that a soil specimen of finite size under external loading is in a uniform state of stress and strain. This assumption is usually made for the interpretation of actual test results. Furthermore it will be assumed that strains are small, so that stresses are proportional to applied loads and strains to displacements. The analysis of the load-displacement behavior for a specimen of finite size is therefore equivalent to that of the stress-strain relationship at a constitutive level.

In matrix terms, the latter can be expressed as:

$$\dot{\sigma}' = D\dot{\varepsilon} \tag{4}$$

Under stress (load) control, unlimited strain (displacement) rates occur when

$$\det D = 0 \tag{5}$$

Equation 5 gives what is usually considered as the failure condition and, for a strain hardening material, the state for which this occurs is characterized by the nullity of the hardening modulus. The locus in the stress space for which Equation 5 is fulfilled can be also considered as a limit locus, since a virgin soil cannot bear a stress state beyond it.

In geotechnical tests, however, usually the controlling parameters are some stress and some strain components. The stress-strain relationship can be therefore rearranged, in order to have on the l.h.s. all the controlling parameters, for instance:

$$\begin{Bmatrix} \dot{\sigma}'_1 \\ \dot{\varepsilon}_2 \end{Bmatrix} = \begin{bmatrix} A_{11} & A_{12} \\ A_{21} & A_{22} \end{bmatrix} \begin{Bmatrix} \dot{\varepsilon}_1 \\ \dot{\sigma}'_2 \end{Bmatrix} \tag{6}$$

where $\dot{\sigma}'_i$ and $\dot{\varepsilon}_j$ are vectors of n and $6-n$ components, respectively, (with $0 \le n \le 6$), while A_{ij} are submatrices that are obtained by rearranging the stiffness matrix D. In this case, the condition for the occurrence of unlimited strains (and stresses) is clearly

$$\det A = 0 \tag{7}$$

It can be shown (Nova (1989), (1994)) that this occurs when

$$\det D_{11} = 0 \tag{8}$$

where D_{11} is the submatrix connecting $\dot{\sigma}'_1$ to $\dot{\varepsilon}_1$. In other words, whenever a minor of the stiffness matrix is nil, there exists a particular set of controlling parameters for which the complementary parameters of stress and strain can increase indefinitely.

Since, in many instances, strain and stress rates are not controlled directly, but the value of a linear combination of them is imposed (generalized stresses and strains, such as volumetric strain or deviator stress), Equation 8 can be generalized to

$$\det \Delta_{11} = 0 \tag{9}$$

where Δ is a matrix connecting generalized stress and strain rate vectors.

It can be shown further that if the stiffness matrix is positive definite, neither one of Equations 5, 8 and 9 can be fulfilled. Unlimited strain rates cannot take place under any loading program. On the other hand, if the stiffness matrix is positive semi-definite, there exists one particular loading program, in terms of generalized stress-strain, for which loss of control occurs. For stress states beyond that level, there is an entire cone of possible load paths for which such a possibility exists (Imposimato and Nova (2001)).

The stiffness matrix of a material obeying an associate flow rule is symmetric and becomes positive semi-definite when the limit locus is achieved. All types of load or displacement control path, even if generalized, give rise to limited responses for small perturbations below that locus. No failure of the type discussed so far is possible, therefore, in the hardening regime ($H > 0$). If the flow rule is non-associate, however, the stiffness matrix is not symmetric, the loss of positive definiteness occurs in the hardening regime and load programs for which unlimited strains occur within the usually considered safe domain are possible (Nova (1989), (1994)).

The occurrence of such failures in the hardening regime is indeed a clue of the fact that the flow rule is non-associate.

Whenever either one of Equations 5,8 or 9 is fulfilled, infinitely many eigen-solutions are possible. On the one hand this implies that the fulfillment of any of these Equations corresponds to a state for which lack of uniqueness of the incremental response takes place. On the other one, this also implies that solutions are possible only in the case load increments are given in a very special way. If not, no solution exists. For instance, when Equation 8 is fulfilled, it is possible an infinity of eigen-solutions of the type

$$\dot{\sigma}_2' = \beta D_{21}\dot{\varepsilon}_1^* + D_{22}\dot{\varepsilon}_2 \tag{10}$$

where β is an indefinite scalar and $\dot{\varepsilon}_1^*$ is the eigen-vector of matrix D_{11}, when the load increment is such that

$$\dot{\sigma}_1' = D_{12}\dot{\varepsilon}_2 \tag{11}$$

This condition is similar to that occurring when a 'latent instability' analysis of structures under load is made (Ziegler (1968)). In that case, infinitely many configurations of the structure are possible, under the same 'critical' load. By analogy, we shall denote the aforementioned conditions as 'latent instability' conditions.

3 EXAMPLES OF LOSS OF CONTROL IN THE HARDENING REGIME

The occurrence of a peak in the undrained test of Fig 2 can be easily predicted in this framework. In an axisymmetric test, the constitutive law can be condensed as follows:

$$\begin{Bmatrix} \dot{\varepsilon}_v \\ \dot{\varepsilon}_d \end{Bmatrix} = \begin{bmatrix} C_{pp} & C_{pq} \\ C_{qp} & C_{qq} \end{bmatrix} \begin{Bmatrix} \dot{p}' \\ \dot{q} \end{Bmatrix} \tag{12}$$

where ε_v and ε_d denote the volumetric and the deviatoric strain, respectively, defined as:

$$\varepsilon_v = \varepsilon_1 + 2\varepsilon_3 \tag{13}$$

$$\varepsilon_d = \frac{2}{3}(\varepsilon_1 - \varepsilon_3) \tag{14}$$

and C_{ij} are the elements of the compliance matrix.

In the case considered, the controlling variables are the volumetric strain, that must be constant, and the axial load. At stress deviator peak, the volumetric compliance C_{pp} must be zero, to comply with the first of Equations 12. At peak, the stress strain relationship is therefore such that:

$$\begin{Bmatrix} 0 \\ 0 \end{Bmatrix} = \begin{bmatrix} -C_{pq}C_{qq}^{-1}C_{qp} & C_{pq}C_{qq}^{-1} \\ -C_{qq}^{-1}C_{qp} & C_{qq}^{-1} \end{bmatrix} \begin{Bmatrix} \dot{p}' \\ \dot{\varepsilon}_d \end{Bmatrix} \tag{15}$$

The determinant of the matrix of Equation 15 is zero and unlimited deviatoric strain and pore pressure rates, $\dot{u}$, can take place:

$$\dot{p}' = -\dot{u} = -\varphi \tag{16}$$

$$\dot{\varepsilon}_d = -C_{qp}\varphi \qquad\qquad (17)$$

where φ can be any (positive) scalar.
  Conversely Equation 18:

$$C_{pp} = C_{ijhk}\delta_{ij}\delta_{hk} = 0 \qquad\qquad (18)$$

gives the condition for a peak of the deviator stress and consequently for latent instability (loss of control) of a load controlled test. It can be shown that Equation 18 is the equation of a straight line passing through the origin of axes in the p',q plane, known as Lade instability line (Lade (1992)).

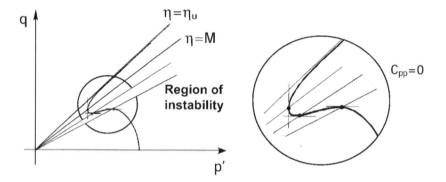

Figure 3. Region of instability in the 'triaxial' plane under load control in undrained conditions

  It is perhaps of some interest to note in passing that the region of instability is of limited extension, as shown in Figure 3 for a medium dense sand. This region is in fact confined by two straight lines, passing through the origin. Both fulfill Equation 18.

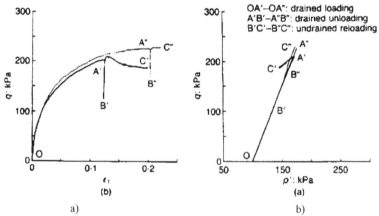

a)   b)

Figure 4. a) stress paths and b) stress-strain relationships for loose Hostun sand specimens preloaded in drained conditions, unloaded and reloaded in undrained conditions. After di Prisco et al. (1995)

  If the stress path by-passes such a region, by drained preloading, for instance, no control loss occurs in load controlled undrained reloading. Figure 4 shows in fact the experimental results conducted on two theoretically identical specimens of loose Hostun sand (di Prisco et al.

(1995)). They differ in that the drained preloading is larger for the second case (OA'') than in the first one (OA'), A' being inside the instability region, while A'' is above it. It is apparent from Figure 4 that the undrained reloading branches show a very different behaviour. A peak of the deviator stress is observed for path B'C', implying instability under load control, while a monotonic increase of the deviator stress is recorded along path B''C''.

Another case of loss of control occurs when the controlling quantities are one principal strain rate, say $\dot{\varepsilon}_1$, and two principal stress rates, say $\dot{\sigma}_2'$ and $\dot{\sigma}_3'$. By grouping the last two terms and the corresponding strain rates in vectors $\dot{\sigma}_\alpha'$ and $\dot{\varepsilon}_\alpha$, respectively, after variable rearrangement, provided the stiffness D_{11} is not nil, the constitutive law can be rewritten as

$$\begin{Bmatrix} \dot{\varepsilon}_1 \\ \dot{\sigma}_\alpha' \end{Bmatrix} = \frac{1}{D_{11}} \begin{bmatrix} 1 & -D_{1\alpha} \\ D_{\alpha 1} & D_{\alpha\alpha}D_{11} - D_{\alpha 1}D_{1\alpha} \end{bmatrix} \begin{Bmatrix} \dot{\sigma}_1' \\ \dot{\varepsilon}_\alpha \end{Bmatrix} \tag{19}$$

According to Schur (1917) theorem, when

$$\det D_{\alpha\alpha} = 0 \tag{20}$$

the determinant of the matrix of Equation 12 is zero. Therefore, in general, the incremental solution does not exist and we cannot assign arbitrarily the increments at the l.h.s. Infinite eigensolutions exist however, for the following load program

$$\dot{\sigma}_\alpha' = D_{\alpha 1}\dot{\varepsilon}_1 \tag{21}$$

Infinite solutions are therefore associated to the same load increment and a homogeneous bifurcation takes place. Note further that it can be shown that, when Equation 8 is fulfilled, also the complementary coefficient of the compliance matrix C_{11} is zero.

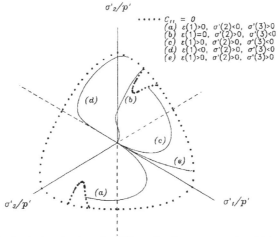

Figure 5. Stress paths in the deviatoric plane for different loading programs in which one principal strain rate and two principal stress rates are controlled. Homogeneous bifurcation loci are starred (after Imposimato and Nova (1998))

For instance, for a convenient set of parameters of the constitutive model 'Sinfonietta Classica' (Nova (1988)), the starred loci for which Equation 13 is satisfied are shown in Figure 5. It is also shown that out of the five stress paths considered, two touch the loci. A loss of control

trol and homogeneous bifurcation occur therefore at such points, under those particular loading programs.

The framework of this analysis can be used also for determining the loci for which a drained shear band takes place (Nova (1989), Imposimato and Nova (1998)). It can be proven that, for an isotropic material, in the principal stress space, such loci coincide with those shown in Figure 5 (plus the symmetric loci obtained by permutation of indices). At the same stress point, therefore, both homogeneous and inhomogeneous (shear band) bifurcation can take place. The difference is in that homogeneous bifurcation is possible only under the very special load paths illustrated in Figure 5, whilst a shear band can occur for any type of load increment, because two different fields of strain and stress rates can coexist *en echelon* violating neither equilibrium nor compatibility. For this reason, in practice, in plane strain problems shear bands are more frequently observed than homogeneous bifurcation.

4 SOFTENING AND LOSS OF CONTROL IN OEDOMETRIC TESTS

Loss of control can occur, in special conditions, even in the most traditional of the geotechnical tests: the confined compression, so called oedometric, test. It will be shown in fact, by means of an elastoplastic strainhardening constitutive model, that during this test, an overconsolidated or a cemented soil specimen can reach the limit condition, soften and then harden again, provided the initial overconsolidation or degree of cementation are large enough. Furthermore, it will be shown that, for a particular choice of the constitutive parameters, appropriate for strong but brittle intergranular bonding and large initial porosity, vertical collapse of the specimen can take place, with possible formation of compaction bands and loss of load control.

Consider first the model for a cemented soil or a soft rock (Nova (1992)) where the bond strength is characterized by two fixed parameters p_t and p_m. The size of the elastic domain is controlled by p_s, that is a hardening variable, depending on the plastic strains experienced by the material. A picture of the initial elastic domain is shown in Figure 6. Assume, for the sake of simplicity, that the flow rule is associated and that hardening depends only on volumetric strains, as in Cam Clay (Schofield and Wroth (1968)). Compaction therefore corresponds to an increase in p_s and conversely dilation to its decrease.

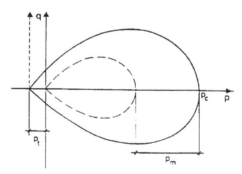

Figure 6. Initial elastic domain for a cemented material – after Nova (1992)

A typical stress path experimentally obtained in an oedometric test on cemented geomaterials, such as chalk (Leddra (1988)) or calcarenite (Coop and Atkinson (1993)), is shown in Figure 7a. The model predicts a similar trend. At the beginning the behavior is elastic, since the origin belongs to the initial elastic domain. The stress path is therefore given by a straight line, whose slope depends on the Poisson's ratio, only. The yield locus is reached at a point for which the horizontal component of the vector normal to it is directed in the negative sense of the hydrostatic axis, see Figure 7b. Plastic dilation is predicted, therefore, and consequently, for the assumed hardening law, softening of the material occurs.

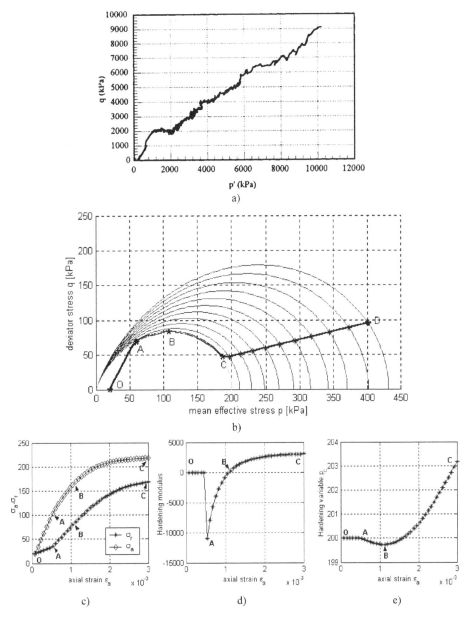

Figure 7. Oedometric test on a cemented geomaterial a) experimental (calcarenite) stress path- after Coop and Atkinson b) Calculated stress path c) stress-strain law d) value of the hardening modulus e)value of the hardening variable –After Castellanza (2002)

From the standpoint of the theory of plasticity, the specimen 'fails' at this point. The stress path is directed within the initial elastic domain and the hardening variable decreases together with its size, Figure 7e. The deviator stress increases, however, as well as the external loading. Full controllability of the test is possible, therefore, even under load control.

It is quite interesting to note that, despite the deviator stress increases, the hardening modulus is negative, so that softening occurs, Fig 7d. On the other hand, softening does not imply loss of controllability of the test, for the special loading program we are following, since vertical stress increases monotonically, Figure 7c. On the contrary, if the load program would be stopped at any moment between points A and B of Figure 7 and a different load increment, say an increase in vertical load at constant horizontal stress, would be imposed, the specimen would instantaneously collapse.

As point B is reached, the normal to the yield locus is directed towards the positive sign of the isotropic axis. Plastic compression takes place as well as hardening, for the assumed hardening and plastic flow rules. Not withstanding that, the deviator stress decreases. The stress path is obliged to follow more or less the shape of the yield locus, since the elastic strains are small and the zero lateral strain condition imposes that the plastic ones are of the same order of magnitude. Only when it is achieved a stress condition for which the normality rule determines plastic strain rates that are compatible with the oedometric condition, strains can occur freely, the stress path can leave the yield locus shape and the deviator stress can increase again.

Remarkably, similar results can be obtained by using a basic model as the original Cam Clay for a preconsolidated material with an overconsolidation ratio high enough to render the specimen 'dryer than critical', i.e. heavily overconsolidated.

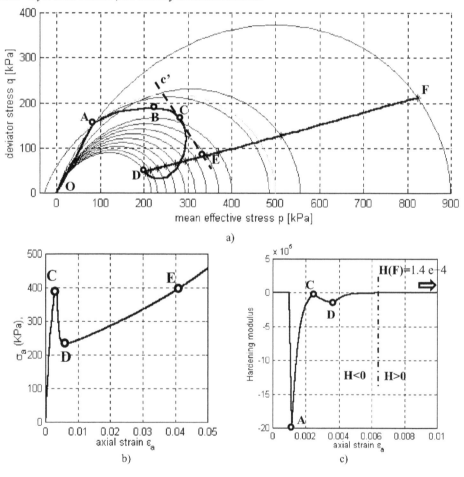

Figure 8. Oedometric test on a cemented geomaterial with bond degradation a) Calculated stress path b) axial stress strain relationship c)value of the hardening modulus –After Castellanza (2002)

Consider now a material whose bonds are brittle. Yielding is therefore associated to a decrease of the value of the variables p_t and p_m. In this case it is possible to have loss of load control even in the oedometric test. Figure 8 shows in fact the calculated results for a material characterized by convenient constitutive parameters. In this case the hardening parameter is given by the sum of three contributions (Gens and Nova (1993)):

$$H = H_s + H_t + H_m \qquad (22)$$

each of which is related to the corresponding hardening variable p_s, p_t, p_m. While the first contribution can be either positive or negative, the other two are always negative. Softening can occur even in the region where the material is compacting, therefore, Figure 8c. The axial stress strain law shows a marked peak, implying that load control is impossible, beyond the peak. The calculated results were in fact obtained by imposing full kinematic control.

In such a test, in a specimen of finite size, but uniform state of stress and strain, at peak a bifurcation can take place without violating equilibrium nor compatibility. The specimen can be seen in fact as composed of layers, which are subject to the same vertical stress decrease but different horizontal stress variation, one layer undergoing elastic unloading while the adjacent one suffers elastoplastic loading, as shown in Figure 9. Such layers have been actually observed in soft rock specimens under compression (Olsson (1999)). Rudnicki (2002), following the same path of reasoning of Rudnicki and Rice (1975) for shear bands, has given the mathematical condition for the occurrence of compaction bands, i.e

$$D_{ijhk} n_i n_j n_h n_k = 0 \qquad (23)$$

where D_{ijhk} is the stiffness tensor and n_i are the director cosines with respect to the reference axes of the normal n to the plane of the band.

According to tensor transformation rules, the l.h.s. of Equation 16 gives the stiffness coefficient D_{nnnn}. Equation 16 becomes therefore:

$$D_{nnnn} = 0 \qquad (24)$$

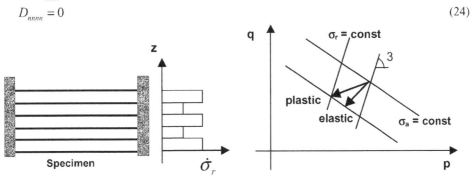

Figure 9. Bifurcation in an oedometric test on a cemented geomaterial with bond degradation a) compaction bands b) stress rates in the layers –After Nova et al. (2003)

The nullity of this stiffness coefficient, that coincides in this case with the occurrence of a peak in the axial stress-strain relationship (horizontal compaction band), is therefore the condition for bifurcation.

Softening can occur in unexpected conditions, therefore. As a final example, consider the results of an oedometric test on a specimen of loose sand, shown in Figure 10. During virgin loading, the ratio between horizontal and vertical stresses remains constant, so that the stress path in the plane p', q is a straight line passing through the origin. For moderate unloading, hysteretic unloading-reloading loops are apparent. If the material is fully unloaded, however, the reloading stress path is again a straight line, i.e. the material behaves as if it was virgin anew.

Such a result can be easily explained within the framework of the model. When unloading is moderate, initially the behavior is elastic while only limited plastic strains take place when the yield locus is reached again (for negative values of q). When unloading is large, on the contrary, after the initial elastic phase, large plastic strains take place and the material softens with a consequent reduction of the size of the elastic domain. If the specimen is fully unloaded, the elastic domain shrinks to a point, the material is in the same state as at the beginning of the test and the stress path upon reloading is that typical of a virgin specimen, i.e. a straight line passing through the origin. In a sense, we can say that the specimen 'fails' during unloading in an oedometric test, since softening is activated. This fact has no dramatic consequences, however, since the state of the material is simply driven back to a virgin condition.

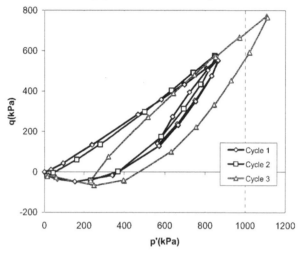

Figure 10. Experimental stress path in an oedometric test on a loose sand specimen, with cycles of unloading-reloading. After Castellanza (2002)

5 CONCLUSIONS

Failure of a material element is a primitive concept. When dealing with geomaterials such a concept becomes less neat, however. Although it is possible to define a convenient limit locus beyond which stress states are not feasible for a given material, in fact, unlimited strains may occur even for stress states within such locus, provided convenient loading conditions are imposed and controlling parameters appropriately chosen.

Such types of failures can occur in the hardening regime, provided the flow rule is non-associated and the stiffness matrix is not positive definite. An example is the loose sand specimen failure in a load controlled undrained test. Although under drained conditions the material is fully stable and shows a typical hardening behavior, if the kinematic and static control parameters are changed (e.g. from axial strain and confining pressure to volumetric strain and axial load), a brutal collapse takes place, for stress levels very far from the limit locus.

Other types of loss of control can occur. Whenever a principal minor of the stiffness (or compliance) matrix is zero, there exists a special combination of stress and strain increments for which an infinity of solutions, in terms of the complementary variables of strain and stress, is possible. Homogeneous bifurcations may take place, therefore.

It is also possible to consider linear combinations of stresses and strains. In this case a generalized stiffness matrix can be defined and a similar conclusion holds in terms of generalized stresses and strains. In particular, it can be shown that even the shear band type of bifurcation can be described within this conceptual framework. In this latter case a non-homogeneous bifurcation is possible, in a specimen of finite size, since neither equilibrium nor compatibility is violated by two different, adjacent, stress-strain rate states.

All the aforementioned bifurcation states may occur even in the hardening regime. The homogeneous bifurcations can take place only if the appropriate control variables are chosen, however. The states for which such bifurcations can occur are 'unstable' (since infinite solutions to a single small perturbation are possible), but such an instability is highlighted only under special load conditions. The situation is similar to that of a mountaineer walking up a steep slope. His state is unstable along the entire path, since he can fall downhill if he stumbles. On the other hand, under appropriate control of his feet, he can quietly reach the mountain top.

Along certain load paths, geomaterials can experience softening. According to the theory of plasticity, this occurs when the hardening modulus is negative. Usually to the term 'softening' are associated the concepts of 'failure' and 'instability'. This is certainly the case when load controlled tests are considered. In general, however, softening is not necessarily associated to specimen collapse and loss of control. For instance, if an oedometric test on a cemented soil is considered, it was shown in the paper that, at variance with simple mind expectations, softening is associated to an increase in deviator stress. On the contrary, hardening is initially associated to a decrease in the deviator stress. The axial load increases in both cases, however, so that no loss of control occurs, even in partially load controlled tests.

If the soil bonds degrade with increasing plastic strains, as it can be the case with calcareous materials, for example, the interplay of hardening and softening is complex. The latter, linked to bond degradation, can prevail on the former, that is due to plastic volumetric compaction. Overall softening can take place together with volumetric compaction, therefore, and a peak in the axial stress-strain curve is recorded in fully kinematically controlled oedometric tests. As a consequence, in a test in which the axial load is controlled, sudden collapse of the specimen (i.e. loss of control) occurs at the stress level corresponding to the peak. Since the peak condition is associated to the nullity of the stiffness coefficient in the axial stress direction, a planar compaction band in the orthogonal direction can be formed. The stress strain state in the oedometric specimen can undergo a non-homogeneous bifurcation, with layers that are elastically unloaded and adjacent layers experiencing plastic strains.

Finally, it has been shown that softening can occur in the oedometric test even for a non cemented geomaterial, when this is fully unloaded. The softening experienced erases from the specimen memory all the previous loading history and the material behaves as it would be fully virgin.

Failure and the related concepts of hardening and softening are by no means intuitive, therefore. Instabilities leading to large strains can occur in the hardening regime, while softening can be associated to perfectly controllable situations. Sudden collapses are possible in oedometric tests and softening can occur in unloading, regenerating virgin soil specimens. A conceptual framework, as the theory of strainhardening plasticity, is necessary in order to fully appreciate and control the complexity of the behavior of geomaterials.

REFERENCES

Castellanza, R. 2002. *Weathering effects on the mechanical behaviour of bonded geomaterials: an experimental, theoretical and numerical study*. PhD Thesis, Politecnico di Milano.

Coop, M.R. & Atkinson, J.H. 1993. The mechanics of cemented carbonate sands. *Géotechnique* 43(1):53-67.

Di Prisco, C., Matiotti, R. & Nova, R. 1995. Theoretical investigation of the undrained stability of shallow submerged slopes. Géotechnique, 45(3): 479-496

Gens, A. & Nova, R. 1993. Conceptual bases for a constitutive model for bonded soils and weak rocks. In A. Anagnostopoulos, F. Schlosser, N. Kalteziotis & R. Frank (eds) *Geotechnical Engineering of Hard Soils-Soft Rocks; Proc. Intern. Symp. Athens*: 485-494. Rotterdam: Balkema

Imposimato, S. & Nova, R. 1998. An investigation on the uniqueness of the incremental response of elastoplastic models for virgin sand. *Mechanics of Cohesive Frictional Materials*, 3(1): 65-87

Imposimato, S. & Nova, R. 2001. On the value of the second order work in homogeneous tests on loose sand specimens. In H.-B. Muehlhaus, A.V. Dyskin & E. Pasternak (eds), *Bifurcation and localisation theory in Geomechanics*: 209-216. Lisse : Swets & Zeitlinger.

Lade, P.V. 1992. Static instability and liquefaction of loose fine sandy slopes. *J. Geotech. Engin. ASCE* 118: 51-71

Leddra, M.J. 1988. *Deformation on chalk through compaction and flow*. PhD Thesis, Univ. of London.

Nova, R. 1988. Sinfonietta classica: an exercise on classical soil modelling. In A. Saada & G. Bianchini (eds) *Constitutive Equations for Granular non-cohesive soils*; *Proc. Intern. Symp., Cleveland*: 501-519. Rotterdam: Balkema

Nova, R. 1989. Liquefaction, stability, bifurcations of soil via strain-hardening plasticity. In E. Dembicki, G.Gudehus & Z. Sikora (eds) *Numerical Methods for the Localisation and Bifurcation of granular bodies; Proc. Int. Works., Gdansk*: TUG 117-132

Nova, R. 1992. Mathematical modelling of natural and engineered geomaterials. General lecture 1st E.C.S.M. Munchen, *Eur.J. Mech. A/Solids*, 11,Special issue: 135-154.

Nova, R. 1994. Controllability of the incremental response of soil specimens subjected to arbitrary loading programmes. *J. Mech. Behav. Mater.* 5(2): 193-201

Nova, R., Castellanza, R. & Tamagnini, C. 2003. A constitutive model for bonded geomaterials subject to mechanical and/or chemical degradation. *Submitted publication*

Olsson, W. A. 1999. Theoretical and experimental investigation of compaction bands in porous rocks. *J. Geophys. Res.* 104: 7219-7228.

Rudnicki J. W. 2002. Conditions for compaction and shear bands in a transversely isptropic material. *Int. J. Solids and Structures* 39: 3741-3756.

Rudnicki, J.W. & Rice, J.R. 1975. Conditions for the localisation of deformation in pressure sensitive dilatant materials. *J. Mech. Phys. Solids* 23: 371-394

Schofield, A.N. & Wroth, C. P. 1968. *Critical State Soil Mechanics*. Chichester: McGraw-Hill.

Schur, I. 1917. Uber Potenzreihen, die im Innern des Einheitskreises beschraenkt sind *Reine Angew. Math.* 147: 205-232

Ziegler, H. 1968. *Principles of structural stability*. Waltham: Blaisdell Publishing Company

Bifurcations & Instabilities in Geomechanics, – Labuz & Drescher (eds.)
© 2003 Swets & Zeitlinger, Lisse, ISBN 90 5809 563 0

On bifurcation and instability due to softening

A.A. Dobroskok, A.M. Linkov
Institute for Problems of Mechanical Engineering (Russian Academy of Sciences), St-Petersburg, Russia

ABSTRACT: The paper focuses on strict definitions of bifurcation and instability applicable to problems involving dynamic phenomena in rock (seismic events, rockbursts, earthquakes). It is noted that in these problems softening is the only physical process, which may cause instability and bifurcation when geometrical non-linearity is inessential or neglected. Discrete models with one and two degrees of freedom (DOF) are considered to clarify the matter. Strict definitions and conditions of instability and bifurcation are given for systems with finite number of DOF. Comments are made on cases when the effects of instability and bifurcation on the level of microstructure manifest themselves macroscopically as instability of crack propagation and bifurcation of its trajectory.

1 INTRODUCTION

Bifurcation and instability are two important effects arising due to geometrical or/and physical nonlinearity. These effects are quite tricky what appears both in confusing terminology when bifurcation is identified with instability and in restrictions in analysis when instability is studied without notion on bifurcation. Hence, when discussing the effects, we must agree about definitions. To do this, it is useful to consider the simplest examples, which allow us to clearly distinguish between the effects. In Section 2 we present such examples from three different fields of solid mechanics and show that they are equivalent in the mathematical sense.

They serve us in Section 3 to suggest definitions, which look appropriate at least for systems with softening elements. The definitions distinguish between non-uniqueness, bifurcation and instability. A rigorous algebraic criterion of instability follows from the accepted definition for systems with a finite number of degrees of freedom (DOF).

More difficult question concerns with a choice of the equilibrium path at a point of bifurcation. Various hypotheses have been suggested to solve it. We compare some of them in Section 4. It appears that sometimes the suggested hypotheses contradict each other. A conclusion on the most plausible one is drawn. Numerical implications are discussed in Section 5.

The third question addressed in the paper is how bifurcation on the level of softening elements manifests itself on the larger level, at which only macroscopic characteristics are available. The subject is discussed in Section 6 for the example of a crack with a small zone of softening material ahead of its tips. We show that the conditions of instability and bifurcation, which govern these effects, are the same on the both levels.

Brief summary concludes the exposition.

2 EXAMPLES AND ANALYSIS OF SYSTEMS WITH ONE AND TWO DOF

We start with detailed analysis of systems, which being sufficiently simple retain important features of more complicated systems. The simplest system is a system with one degree of freedom (Fig. 1a). It is represented by a rod with a diagram having a descending portion with the softening slope N. The rod is loaded through a spring with rigidity C. This scheme was suggested by Cook (1965) to explain and to study rockbursts in mines. The rod represents a softening pillar, the spring represents elastic embedding rock.

It is well-known, that the rod after the maximum load P_o is fractured violently if the softening slope N is greater than the rigidity of the spring C. The energy excess, transformed into kinetic energy of fractured fragments, is shown by the dashed area in Figure 1b. The violent fracture is associated with instability. The condition of instability is $N \geq C$. The system exhibits instability, but being geometrically linear and having only one DOF, it cannot exhibit branching.

The next in complexity after the systems with one DOF are systems with two DOF. They have new important properties usual in systems with greater DOF. In particular, they exhibit possibility of branching in their behavior. Consider three examples given by authors in different fields of solid mechanics.

The first of them (Fig. 2) refers to "the column paradox", discussed by F. Shanley (1946, 1947). His paper of 1947 is considered "epoch-making" (Bazant & Cedolin 1991). The paper tends to study instability of a geometrically nonlinear system of two columns, which after elastic deformation with the module E may experience plastic hardening with the tangent module E'. The columns are loaded by the axial force P through rigid rods. The increment of the load is δP. Following Klushnikov (1980), and Bazant & Cedolin (1991), we introduce also an increment of the horizontal force δQ. Finally, we obtain the system of equations describing the behavior of the system after the yielding limit is reached in both rods:

$$-\delta\varphi = R\delta\lambda - \delta f, \qquad \delta\varphi \leq 0, \qquad \delta\lambda \geq 0 \qquad \delta\varphi^T \delta\lambda = 0, \tag{1}$$

where $\mathbf{R}$ is the main matrix, which completely defines the behavior of the system:

$$\mathbf{R} = \begin{pmatrix} a & 1 \\ 1 & a \end{pmatrix}, \tag{2}$$

$$a = \frac{2}{1 - E'/E} \frac{P_e - P}{P} - \frac{2P_e - P}{P} \frac{1}{E},$$

$P_e = FEh^2/(ld)$ is the Euler's critical load for ideally elastic columns with geometrical parameters h, l and d clear from Figure 2, F is the area of the total cross section of the rods;

$$\delta\varphi^T = (\delta\varphi_1, \delta\varphi_2), \qquad \delta\lambda^T = (\delta\lambda_1, \delta\lambda_2), \qquad \delta f^T = (\delta f_1, \delta f_2);$$

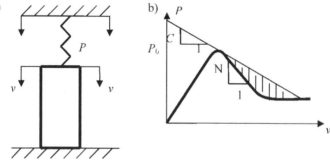

Figure 1. Cook's scheme explaining dynamic fracture in a mine.

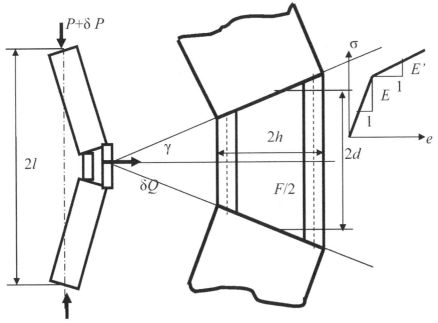

Figure 2. Shanley's scheme of two columns experiencing plastic hardening deformations.

$$\delta\varphi_i = 2\frac{P_e - P}{P}\frac{1}{E}\delta\varphi_{0i}$$

is the normalized increment of the non-positive function $\varphi_{0i} = \sigma_i - (\sigma_0 + E'\Delta\lambda_i)$ defining the yielding surface $\varphi_{0i} = 0$ of the i-th element ($i = 1, 2$); $\delta\lambda_i$ is a non-negative plastic multiplier (in the considered problem it is equal to the absolute value of an increment of plastic strain in the i-th column); δf_i are the components of the generalized external loads:

$$\delta f_1 = 2\left(\frac{P_e - P}{P}\delta p + \frac{P_e}{P}\delta q\right), \qquad \delta f_2 = 2\left(\frac{P_e - P}{P}\delta p - \frac{P_e}{P}\delta q\right),$$

$\delta p = \delta P/(EF)$, $\delta q = \delta Qh/(2P_e d)$; the superscript T denotes transposition.

Emphasize two important features of the system (1) having roots in the physical nature of the discussed problem. Firstly, the main matrix **R** is multiplied by the vector $\boldsymbol{\delta\lambda}$ of plastic multipliers, which cannot be negative. This reflects irreversible nature of plastic deformations. Secondly, the system is not linear because its last equation contains the product of the unknown vectors $\boldsymbol{\delta\lambda}$ and $\boldsymbol{\delta\varphi}$. This reflects the fact that at a limit state there is an alternative: an element may experience either (i) further plastic deformation or (ii) elastic unloading. The problem (1) is called the problem of linear complimentarity.

The second example is given by Maier with co-authors (1973). It refers to a beam with two plastic hinges at the points with the greatest moments, where external forces $P_1 = P_2 = P$ are applied (Fig. 3a). A hinge i ($i = 1, 2$) exhibits linear softening with the negative modulus H_i after the moment M_i in its cross section reaches the maximum value M_0 (Fig. 3b). Maier et al. (1973) assumed that increments of the forces are equal ($\delta P_1 = \delta P_2 = \delta P$). For further discussion, we will consider more general case when the increments may be different. After derivations we arrive at the same system (1) with the matrix (2), where now

$$a = 2 + (9/2)Hl/(EJ), \quad \varphi_i = M_i - (M_0 + H\Delta\theta), \quad \delta\lambda = (2/9)(EJ/l)\delta\theta,$$

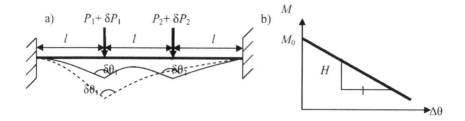

Figure 3. The scheme of a beam with softening hinges (Maier et al. 1973).

$$\delta f_1 = (1/27)l(8\delta P_1 + \delta P_2), \quad \delta f_2 = (1/27)l(\delta P_1 + 8\delta P_2),$$

$3l$ is the length of the beam, E is its elastic module, J is the moment of inertia of the cross section; $\Delta\theta_i$ is the total angle of plastic rotation in the i-th hinge, $\delta\theta_i$ is an increment of this angle.

The third example (Fig. 4) refers to a system of two softening pillars at adjacent seams (Linkov 1997). It can be also interpreted in terms of plastic necks between cracks. Pillars may experience softening with the negative modulus H after the loads acting on them reach the maximum value P_0. By representing a pillar as a boundary element with constant displacement discontinuity, we again arrive at the system (1) (Linkov 1994, p.45-48). Its matrix $\mathbf{R}$ can be written in the form (2), where now $a = (H + z_{11})/z_{12}$, z_{ij} are the coefficients of the distortion matrix, which in the considered case is inverse to the compliance matrix of embedding rocks. For the symmetric scheme of Figure 4, we have $z_{11} = z_{22} > 0$, $z_{12} = z_{21} > 0$; $\delta\lambda_i = \delta v_i^P/z_{12} \geq 0$; δv_i^P is the plastic deformation of the i-th pillar ($i = 1, 2$); δf_1, δf_2 are normalized increments of external loads arising due to mining.

Note that in each of the three examples the coefficient a in the main matrix $\mathbf{R}$ may be any real value. In particular, it may become less than unit and even negative for sufficiently large external load P in the Shanley's example with *geometrical* non-linearity (Fig. 2) or for sufficiently large negative slope $|H|$ for the examples with specific physical nonlinearity, that is softening (Fig. 3, 4). We see that in the mathematical form, the three problems are equivalent and any of them may serve to study specific behavior of a system with irreversible deformations.

For certainty, we shall have in mind the scheme of the beam with softening hinges. This problem was comprehensively studied in (Maier et al. 1973) for the case of equal increments $\delta f_1 = \delta f_2$ of external loads. We will use some results of this analysis and complement them with the discussion of the case $\delta f_1 \neq \delta f_2$, which will enable us to conclude on non-symmetric systems.

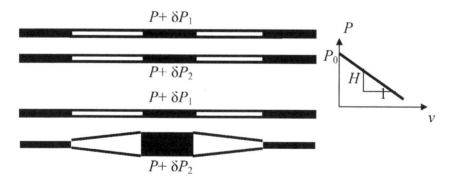

Figure 4. The scheme of two softening pillars at adjacent seams (Linkov 1997).

20

2.1 Non-uniqueness of a solution

Consider the case $a = 1$. Then if $\delta f_1 = \delta f_2 > 0$, there exists only one branch of solutions of (1):

$$\delta\varphi_1 = 0,\ \delta\varphi_2 = 0,\ 0 \le \forall \delta\lambda_1 \le \delta f_1,\ \delta\lambda_2 = \delta f_1 - \delta\lambda_1,$$

but this branch contains infinite number of solutions. They correspond to active softening of both hinges with arbitrary plastic deformation $\delta\lambda_1$ satisfying inequalities $0 \le \delta\lambda_1 \le \delta f_1$. Emphasize that all the solutions are restricted: they tend to zero when the external action δf_1 tends to zero.

This case shows that there may be infinite number of solutions along a single branch. Thus, we need to distinguish between non-uniqueness and bifurcation.

2.2 Branching

Consider the case $0 < a < 1$ when $\delta f_1 > \delta f_2 > 0$, while $\delta f_2 > a \delta f_1$. Analysis shows that there are three branches in this case:

Branch 1 corresponding to softening of the first element and elastic unloading of the second:

$$\Delta\varphi_1 = 0,\ \Delta\varphi_2 = \delta f_2 - \delta f_1/a,\ \delta\lambda_1 = \delta f_1/a,\ \delta\lambda_2 = 0;$$

Branch 2 corresponding to softening of the second element and elastic unloading of the first:

$$\Delta\varphi_1 = \delta f_1 - \delta f_2/a,\ \Delta\varphi_2 = 0,\ \delta\lambda_1 = 0,\ \delta\lambda_2 = \delta f_2/a;$$

Branch 3 corresponding to softening of both elements:

$$\Delta\varphi_1 = 0,\ \Delta\varphi_2 = 0,\ \delta\lambda_1 = (\delta f_2 - a\ \delta f_1)/(1 - a^2),\ \delta\lambda_2 = (\delta f_1 - a\delta f_2)/(1 - a^2).$$

It is quite clear that in the considered case small perturbation of the main matrix $\mathbf{R}$, which makes it non-symmetric, does not exclude any if these branches. This shows that branching is possible in systems without any symmetry.

2.3 Second order work

The quadratic

$$I = \mathbf{x}^T \mathbf{R}\, \mathbf{x} \tag{3}$$

is equal to the normalized second order work ΔW (Maier et al 1973): $\Delta W = I$. Since the vector $\delta\lambda$ entering (1) is non-negative, we conclude that in the considered problems, the second order work is positive if and only if the main matrix is strictly copositive that is:

$$\mathbf{x}^T \mathbf{R}\, \mathbf{x} > 0 \qquad\qquad \text{for } \forall \mathbf{x} \ge \mathbf{0}, \qquad\qquad \mathbf{x} \ne \mathbf{0}. \tag{4}$$

For the matrix $\mathbf{R}$ of the form (2), it is strictly copositive when $0 < a < 1$.

2.4 Second order work for various branches

Consider the same case $0 < a < 1$ when $\delta f_1 > \delta f_2 > 0$, while $\delta f_2 > a\delta f_1$. By using (3) for each of the branches, discussed above, we obtain:

$$\Delta W_1 = \delta f_1^2/a,\ \Delta W_2 = \delta f_2^2/a,\ \Delta W_3 = (-a\delta f_1 - a\delta f_2 + 2\delta f_1\delta f_2)/(1 - a^2).$$

From the analysis of these expressions we conclude that in the considered case

$$0 < \Delta W_3 < \Delta W_2 < \Delta W_1.$$

This means that the second order work ΔW, being positive, is minimum on the branch 3, which corresponds to active plastic loading of the both elements. The branches 1 and 2 with unloading of one of the elements require greater energy expenditure. We conclude that the path with unloading may not coincide with the branch with the minimum second order work. Furthermore, in the considered example, the path 1 with unloading of the second element corresponds to the maximal second order work.

3 NON-UNIQUENESS, BIFURCATION AND INSTABILITY

The examples of the previous section show that in problems involving violent fracture and branching, it is reasonable to distinguish between non-uniqueness, bifurcation and instability. We will use the following definitions for a system having an arbitrary finite number of DOF and containing elements which in a limit state may experience either plastic deformations or elastic unloading. The problem is reduced to the quasi-linear system of equations (1) with the matrix **R** and vectors $\delta\lambda$ and $\delta\varphi$ of order n.

3.1 *Non-uniqueness.*

The definition of *non-uniqueness* is trivial: there exist more than one solutions of the problem (1). This number may be infinite as it was in the case 2.1 above.

3.2 *Bifurcation*

We consider solutions $\delta\varphi^1$, $\delta\lambda^1$ and $\delta\varphi^2$, $\delta\lambda^2$ of (1) to belong to the same *branch* if non-zero components of the vector-columns $\delta\varphi^1$ and $\delta\varphi^2$ are located in the same positions. Thus, in physical sense, branches differ by elements which experience unloading. We call *bifurcation* when a system (1) has more than one branch of solutions. A state in which branching occurs is termed the bifurcation point. As stated above for the case 2.1, a branch may include infinite number of solutions. This implies that non-uniqueness may occur without bifurcation.

3.3 *Stability and instability*

Definitions of stability and instability strongly depend on a problem considered. For the 'column paradox', this was clearly expressed by von Karman in his comments to the paper by Shanley (1947, p. 267-268). After noting that the difference between his and Shanley's analysis is due to the different definitions of instability, von Karman wrote: "My original analysis ... is a generalization of the reasoning used in the theory of elastic buckling. Why does this not cover all possible equilibrium positions in the inelastic case? Obviously, it is... because of the nonreversible character of the process. ... Hence, the definition of the stability must be revised for nonreversible processes. This necessity was intuitively recognized by Mr. Shanley, which is, I believe, a great merit of his paper."

Actually, Shanley identified the bifurcation of the equilibrium path with instability. This was natural for a specialist who worked at the Lockheed Aircraft Corporation: Shanley clearly explained why his approach was reasonable for aluminum constructions.

Meanwhile, other authors solving other problems do not associate bifurcation with instability. They use term "stable bifurcation" (Bazant & Cedolin 1991, Maier & Perego 1992) for the same solution of equations (1), which Shanley termed instability. The arguments for the alternative definition are:
– only infinitesimal deformations correspond to infinitesimal changes of external loads in this case and
– there is no energy excess and consequently no dynamic effects occur.

This paper, in accordance with its title, focuses on systems with softening elements. Having in mind geomechanical applications, we prefer to follow Cook (1965) and associate instability only with violent fracture. Then bifurcation becomes an effect distinct from instability: it may be stable as in the case discussed in (Bazant & Cedolin 1991, Maier & Perego 1992), or unstable in other cases.

The association of instability with dynamic effects leads to the general definition of stability (Linkov 1977, 1994): a state is stable if the difference between the increment of work ΔA of external forces and the increment of internal energy of a system (ΔU) is negative for any kinematically admissible virtual displacements:

$$\Delta K = \Delta A - \Delta U < 0. \tag{5}$$

Otherwise, the state is termed unstable. Then there exist some kinematically admissible virtual displacements, for which

$$\Delta K \geq 0. \tag{6}$$

The condition (6) implies that that there is energy excess which will increase velocities of a movement starting at a considered equilibrium state.

In many cases, $\Delta K = -\Delta W$, and the stability condition (5) is equivalent to the condition of positive second order work (see, e. g. Bazant & Cedolin 1991).

Note that the definitions are formulated in terms of perturbations in displacements. In general, the latter may not correspond to some solution of (1) because they presume dynamic rather than static response of a system. Meanwhile, in particular cases, virtual displacements may correspond to a static solution of (1).

The detailed analysis of geometrically and physically nonlinear systems shows that the accepted definition (5) leads to the algebraic condition of stability, which is not only sufficient but, what is much more difficult to prove, is also necessary for stability. This condition is formulated as a theorem of stability (Linkov 1987). In a simplified formulation, relating to geometrically linear systems with softening elements it reads: a system with a finite number of DOF is stable if and only if its main matrix is strictly copositive, that is satisfies (4).

3.4 *Stable and unstable bifurcation*

The accepted definitions distinguish between bifurcation and instability. Indeed, instability may occur in a state which excludes bifurcation. For instance, this is the case in a system with one DOF (Fig. 1).

On the other hand, bifurcation may be stable or unstable. In particular, in the case 2.2 discussed in the previous section for the examples presented in Figures 2-4, we had stable bifurcation: the stability condition (5) was fulfilled for each of the solutions. Moreover, in accordance with the theorem of stability, bifurcation, when it takes place, is stable in the whole interval $0 < a \leq 1$, where the matrix $\mathbf{R}$ is strictly copositive. But if $a \leq 0$, the matrix $\mathbf{R}$ is not strictly copositive. Then in the interval $-1 < a \leq 0$, where the matrix $\mathbf{R}$ being not strictly copositive is yet not negative definite, we still have solutions of (1), but now some of them correspond to unstable bifurcation. For them, the instability condition (6) is fulfilled on corresponding displacements (the latter may be taken as virtual displacements). As shown by Maier et al. (1973) for $\delta f_1 = \delta f_2 < 0$, the paths with softening of one of the elements correspond now to an algebraically lesser second order work than the path with elastic unloading of both elements.

4 CHOICE OF BIFURCATION PATH

At a point of bifurcation we need to choose a branch of solution. How to do this? A theoretical answer cannot be obtained without using an additional hypothesis. Among those mention the following:
- "structures without symmetry (or at least some hidden symmetry) do not exhibit equilibrium path bifurcation (Bazant & Cedolin 1991, p.666);
- "the incremental solutions of minimal second order work" are "those which will actually occur when stability is ensured" (Maier et al. 1973, p. 901);
- the response one should expect is that, which "leads to situation where equilibrium branching cannot show up again" (Maier & Perego 1992, p.337);
- "from possible bifurcation paths, a system chooses that path which leads to unloading of at least some of its elements" (Linkov 1997, p. 154);
- "a simple rule ... consists in taking the path that involves elastic unloading of the maximum number of plastic elements" (Linkov 2001, p. 433];
- a system in a stable state follows the path, for which the internal entropy change is maximum (Bazant & Cedolin 1991, p. 655).

Consider these assumptions. The first of them, as evident from the case 2.2, is not true: bifurcation may occur in a system with no signs of symmetry. We could also see in the case 2.4 that the second and the third hypotheses sometimes contradict each other. One needs to choose between them or reject both of them. In the authors view, the third hypothesis (Maier and Perego

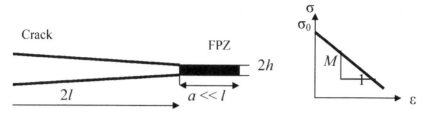

Figure 5. A crack with a thin softening layer at its tip.

1992) is preferable. Consequently, the forth and the fifth hypotheses(Linkov 1997, 2001) present successively stronger formulations of the same idea that a system chooses the way "where equilibrium branching cannot show up again". The sixth hypothesis suggested by Bazant, looks the most promising for stable bifurcation. For the case of controlled loads this suggestion is equivalent to the maximal second order complementary work (Bazant & Cedolin 1991, p.653], what agrees with the hypothesis 3-5 when applied to the example considered in 2.4. Meanwhile, the sixth hypothesis refers only to stable bifurcation. The question of a proper choice of the path for unstable bifurcation is open for further discussion and investigation.

Certainly, any suggestion must be justified by direct physical observations. It looks that the process of localization, when the most part of a system unloads while plastic strain concentrates in a neck or in shear bands, supports the suggestions 3-6.

5 ON NUMERICAL SIMULATION OF BIFURCATION

It is uneasy task to find all the solutions of the problem (1) when the number of elements in the limit state exceeds ten. Even to check whether a main matrix $\mathbf{R}$ of order N is strictly copositive, one needs to check about 2^N algebraic inequalities involving co-factors of the matrix $\mathbf{R}$ (Motzkin 1965, see also Cottle et al. 1970). Thereby, in practical calculations it might be reasonable not to seek all the solutions of (1) to choose between them but rather to rely on random nature of small perturbations arising due to round-off errors and approximate satisfaction of inequalities. Hopefully, the perturbations will automatically lead to the branch, which is statistically preferable. This, at least partly, resembles the choice of a branch in a real physical system that is influenced by random factors. Note that the latter present the only reason, which leads to a particular branch in an absolutely symmetric system such as the Euler's column under the critical load.

Numerical experiments, which simulate localization in systems with many elements in a plastic state, may serve to illustrate the approach (e. g. Cundall 1990). Perhaps, it may be useful to introduce artificial random perturbations into a computer code and to conclude on statistical features of branching.

6 INTERCONNECTION BETWEEN MICROSCOPIC AND MACROSCOPIC PHENOMENA

Nonlinear irreversible process leading to bifurcation or/and instability may be studied on various levels of structure. For any of the levels, one may use the same physical considerations, definitions and criteria of instability and bifurcation. This is the case, for instance, when we apply the condition of instability (6)

$$\Delta K = \Delta A - \Delta U \geq 0 \tag{7}$$

for studying crack propagation.

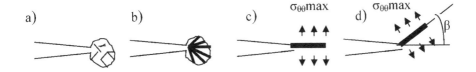

a)   b)   c) $\sigma_{\theta\theta}\text{max}$ d) $\sigma_{\theta\theta}\text{max}$ β

Figure 6. The scheme explaining a connection between bifurcation on micro- and macroscopic level.

When solving the boundary value problem for a crack with a small fracture process zone (FPZ) in a form of a narrow layer of softening material ahead of the crack tip (Fig. 5), we arrive at the following criterion of instability (Petukhov & Linkov, 1979):

$$\frac{1-v^2}{E} K_1^2 \geq \frac{1}{2} \frac{\sigma_0}{M} 2h, \tag{8}$$

where E is the Young's modulus of a medium, K_1 is the stress intensity factor, the meaning of other symbols is clear from Figure 5.

The same condition immediately follows when (6) is applied on the macroscopic level not involving a solution of a boundary value problem. Then (7) actually presents Griffith's criterion, which suggests that the energy release rate $-d\Im/dS$ exceeds or at least equal to energy consumption $2g$ per unit surface:

$$-d\Im/dS \geq 2g. \tag{9}$$

In the considered case we have (see, e. g. Rice 1968) $-d\Im/dS = (1-v^2)K_1^2/E$, while the maximum energy consumption $2g = \sigma_0^2 2h/(2M)$. The result coincides with the condition (7) obtained on the 'microscopic' level. Note that the condition (8) may be written as $K_1 \geq K_{1C}$.

Similar relation, may exist between bifurcation studied on various levels. Consider for example a crack in 2D (Fig. 6a). Recall that crack propagation is an ultimate stage of localization, and let the material be such that the rupture mode under tensile stress is dominant in a small FPZ. This implies that at a point of bifurcation, the system of elements in the FPZ chooses a way corresponding to localization along some of the rays emerging from the tip (Fig. 6b). The actual direction of the ray is normal to the line of maximum tensile stress (max $\sigma_{\theta\theta}$) in the vicinity of the tip. The direction may be tangent to the initial crack (Fig. 6c) or comprise an angle β with it (Fig. 6d) depending on the orientation of max $\sigma_{\theta\theta}$.

Assume that the amount of energy $2g$, which may be consumed in the FPZ during localization and final rupture, does not depend on a particular orientation of max $\sigma_{\theta\theta}$. Then the only parameter governing crack propagation is the energy release rate $-d\Im/dS$ and on the macroscopic level, the crack follows a trajectory along which $-d\Im/dS$ is maximum. For the considered rupture mode, this is the path with maximum positive K_1 or, equivalently (Erdogan & Sih 1963, Cotterell & Rice 1980), with zero K_{II}. Hence, on the macroscopic level, we have the condition defining the external load:

$$K_I = K_{IC}, \tag{10}$$

and the condition defining the direction of the crack propagation:

$$K_{II} = 0. \tag{11}$$

In particular cases there may be a number of rays for which the tensile stress in the FPZ $\sigma_{\theta\theta}$ has local maximum. The preferable one is that with maximum value of $-d\Im/dS$. Macroscopically, this microscopic bifurcation may appear as bifurcation of the crack trajectory: the latter chooses the direction corresponding to the greatest K_1.

Figures 7 and 8 present examples of such bifurcation of a crack trajectory. The first of them refers to bi-axial tension of a plate with a straight crack. For this scheme, the straight trajectory (Fig. 7a), being symmetric, satisfies the condition (11). Meanwhile, the theoretical analysis (Cotterell & Rice 1980, Melin 1983) shows that if $q > p$, there exist branches with $K_{II} = 0$ and

25

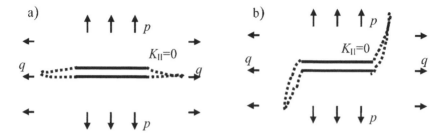

Figure 7. Bifurcation of crack trajectory under bi-axial tension when $q > p$.

greater values of K_1. They correspond, for instance, to the trajectories shown in Figure 7b by dotted lines. Actually, we have bifurcation of the crack trajectory. Experiments (Radon et al. 1977) confirm that for $q > p$ a crack follows the curvilinear branch.

Calculations, performed by using the complex variable boundary element method to find a trajectory satisfying the condition (11), automatically led to the curvilinear branch (Mogilev-skaya 1997, Dobroskok et al. 2001). This agrees with the suggestion of the previous section that rounding-off errors and other sources of random perturbation in numerical calculations lead to the physically significant branch of a solution.

Another example of bifurcation is presented in Figure 8. Again, besides the straight trajectory (Fig. 8a), there exist curvilinear trajectories with $K_{II} = 0$ (Fig. 8b), when crack tips are sufficiently close to each other. Again both the theoretical analysis and experiments (Melin 1983) show that cracks follow curvilinear trajectories of the type of so called overlapping spreading centers (Sempere & Macdonald 1984, Fig. 1a-e). As noted by the latter authors: "The characteristic geometries of overlapping centers is observed over nine orders of magnitude." Again numerical experiments resulted in the curvilinear branches.

7 CONCLUSIONS

The conclusions of the paper are summarized as follows.
- It is reasonable to distinguish between non-uniqueness, bifurcation and instability. For problems connected with violent fracture due to softening (seismic events, rockbursts, earthquakes), we consider appropriate to associate instability with energy excess, while bifurcation with branching.
- Bifurcation may occur in systems without any symmetry.
- A system with softening elements is stable if and only if its main matrix is strictly copositive.
- Stable bifurcation of the equilibrium path is a usual effect in systems with strictly copositive while not positive definite main matrix.

The suggestion by Bazant to choose the branch, which corresponds to the maximal change of internal entropy, looks preferable for stable bifurcation. Still the question of proper choosing between branches is open for further discussion, experimental investigation and detailed analysis.

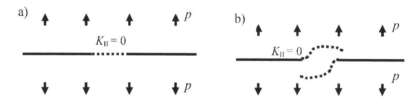

Figure 8. Bifurcation of a crack trajectory in a form of overlapping spreading centers.

– Instability and bifurcation on the level of microstructure may appear as instability and bifurcation on the macroscopic level.

ACKNOWLEDGEMENTS. The authors appreciate the support of the Russian Fund of Fundamental Investigations (Grant 00-05-64316). The first of the authors is also grateful to the St-Petersburg Scientific Council (Grant M01-22K-235). The second of the authors is grateful to the Organizing Committee of the IWBI-2002 and personally to Professor Joseph Labuz for facilitating the participation in the Workshop.

REFERENCES

Bazant, Z.P. & Cedolin, L. 1991. *Stability of Structures: Elastic, Inelastic, Fracture, and Damage Theories*. New York: Oxford University Press.

Cook, N.D.G. 1965. A note on rockbursts considered as a problem of stability. *J. South Afr. Inst. Min. and Metallurgy* 65: 437-446.

Cotterell, B. & Rice, J.R. 1980. Slightly curved or kinked cracks. *Int. J. Fracture* 16: 155-168.

.Cottle, R.W. et al. 1970. On classes of copositive matrices. *Linear Algebra and its Applications* 3: 295-310.

.Cundall, P.A. 1990. Numerical modelling of jointed and faulted rock. In P. Rossmanith (ed.), *Mechanics of Jointed and Faulted Rock*:11-18. Rotterdam: Balkema.

Dobroskok, A.A. et al. 2001. On a new approach in micromechanics of solids and rocks. In Tinucci & Heasley (eds), *Rock Mechanics in the National Interest. Proc. 38th US Rock Mechanics Symposium*: 1185-1190. Rotterdam: Swets & Zeitlinger Lisse.

Erdogan, F. & Sih, G.C. 1963. On the crack extension in plates under plane loading and transverse shear. *Trans. ASME. J. Basic Engineering* 85: 519–527.

Klushnikov, V.D. 1980. *Stability of Elastic-Plastic Systems*. Moscow: Science (In Russian).

Linkov, A.M. 1977. On stability conditions in fracture mechanics. *Proc. Acad. Sci. USSR*: 233 (1): 45-48 (translated into English in "*Soviet Physics. Doklady*").

Linkov, A.M. 1987. Stability of inelastic, geometrically nonlinear, discrete systems. *Soviet Physics. Doklady* 32 (5): 376-378.

Linkov, A.M. 1994. *Dynamic Phenomena in Mines and the Problem of Stability*. Lisboa: Int. Soc. Rock Mech..

Linkov, A.M. 1997. Key-note address: New geomechanical approaches to develop quantitative seismicity. In S. J. Gibowicz & S. Lasocki (eds), *Proc. 4-th Int. Symp. Rockbursts and Seismicity in Mines*: 151-166. Rotterdam: Balkema.

Linkov, A.M. 2001. On modeling of slow and rapid processes in synthetic seismicity. In G. van Aswegen, R. J. Durrheim & W. D. Ortlepp (eds), *Proc. 5th Int. Symposium Rockbursts and Seismicity in Mines, RaSiM-2001*: 433-437. Santon: The South African Institute of Mining and Metallurgy.

Maier, G. & Perego, U. 1992. Effects of softening in elastic-plastic structural dynamics. *Int. J. Numer. Methods Eng.* 34: 319-347.

Maier, G. et al. 1973. Equilibrium branching due to flexural softening. *J. Engineering Mechanics Division* 99: 895-901.

Melin, S. 1983. Why do cracks avoid each other? *Int. J. Fracture* 23: 37-45.

Mogilevskaya, S.G. 1997. Numerical modeling of 2-D smooth crack growth. *Int. J. Fracture* 87: 389-405.

Motzkin, T.S. 1965. Quadratic forms positive for nonnegative variables not all zero. *Notices of American Mathematical Soc.* 12: 224.

Petukhov, I.M. & Linkov, A.M. 1979. The theory of post-failure deformations and the problem of stability in rock mechanics. *Int. J. Rock Mech. Mining Sci. & Geomech. Abstr.* 16 (2): 57-76.

Radon, J.C. et al. 1977. Fracture toughness of PMMA under biaxial stress. *Fracture* 8: 1113-1116.

Rice, J.R. 1968. Mathematical analysis in the mechanics of fracture. In H. Liebovitz (ed.), *Fracture, v. II*: 191-311. New York, London: Academic Press.

Sempere, J.-C. & Macdonald, K. 1984. Overlapping spreading centers: Implications from crack growth simulation by the displacement discontinuity method. *Tectonics* 5: 151-163.

Shanley, F. 1946. The column paradox. *J. Aeronaut. Sci.* 13 (2): 278.

Shanley, F. 1947. Inelastic column theory. *J. Aeronaut. Sci.* 14 (5): 261-268.

Bifurcations & Instabilities in Geomechanics, – Labuz & Drescher (eds.)
© 2003 Swets & Zeitlinger, Lisse, ISBN 90 5809 563 0

Compaction Bands in Porous Rock

J. W. Rudnicki
Department of Mechanical Engineering, Northwestern University, Evanston, Illinois, USA

ABSTRACT: Compaction bands are localized, planar zones of compressed material that form perpendicular to the maximum compressive stress. Compaction bands have been observed in porous sandstone formations in the field and in laboratory compression tests on porous sandstone, plaster, glass beads, polycarbonate honeycombs and metal foams. Because the permeability of the compacted material is much reduced, compaction bands can form barriers to fluid flow and adversely affect attempts to inject or withdraw fluids from porous reservoirs. An analysis of conditions for the onset of compaction bands as a bifurcation from homogeneous deformation indicates that they can occur when the stress strain curve for uniaxial compressive deformation has a peak. Formation of compaction bands is favored by significant compactive inelastic deformation and by a reduction in the yield stress for shear with increasing compressive mean stress. These conditions are typical for stress states on a "cap" yield surface often used to model the inelastic deformation of porous geomaterials.

1 INTRODUCTION

Localized deformation in rocks and other geomaterials has been primarily associated with shear deformation. But Vardoulakis and Sulem (1995, p. 177) note that "compaction layering is observed in geomaterials at many scales." Furthermore, recent field and experimental observations in porous sandstones have identified localized, planar bands of material that has been compacted, without evident shear, relative to the adjacent material. The material in the band is less permeable than the surrounding material (Papamichos et al. 1993) and a decrease of an order of magnitude in permeability has been measured in recent experiments (Holcomb & Olsson, 2002).

Porous rock formations are exploited for a variety of human activities, e.g. recovery and storage of oil and gas, storage of water, disposal of hazardous waste products and sequestration of carbon dioxide to mitigate adverse effects on the atmosphere (Wawersik et al. 2001). Formation of compaction bands caused by stress changes due to the injection or withdrawal of fluid could dramatically alter the fluid flow properties of the formation and prevent efficient, safe and economic use. Compaction band formation is one possible explanation for localized compaction observed in producing oil fields (Nagel, 2001). Furthermore, because compaction bands are localized features, they will be difficult to detect by surface or borehole geophysical methods. Consequently, understanding the conditions that affect the formation and evolution of compaction bands is of practical importance.

Although this paper focuses on compaction bands in geomaterials, similar structures have been observed in a variety of other porous materials, including metal foams (Bastawros et al. 2000, Park & Nutt, 2001), polycarbonate honeycomb (Papka & Kyriakides, 1998), and snow (J. Desrues, pers. comm.). Formation of zones of localized deformation, either compaction or shear, can adversely affect forming processes of metal powders (Gu et al. 2001). Non-uniform density distributions resulting from cold compaction can result in shape distortions during sintering, and final products with undesirable non-uniform properties. Although the micromechanical mechanisms of deformation are different in these various materials, similar phenomenological constitutive relations can be used to describe them (Gu et al. 2001). Thus, understanding the conditions for the occurrence of compaction bands is relevant to a variety of materials and technological processes.

This paper describes in more detail the nature of compaction bands in geomaterials, recent observations of them in laboratory experiments and analyses for their formation. The paper concludes with a discussion of open questions.

2 WHAT ARE COMPACTION BANDS?

Figure 1 shows schematically the formation of a shear band or fault in an axisymmetric compression experiment: relatively homogenous deformation (Fig. 1a) gives way to localized deformation in a narrow planar zone (Fig. 1b). Deformation in the zone is a combination of shear and dilation or compression relative to the plane of the band. The orientation of the band varies with rock type and the confining (lateral) stress but typically makes an angle of about 30° with the direction of the axial compressive stress.

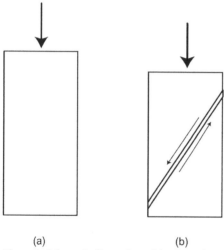

(a)   (b)

Figure 1. Schematic illustration of the formation of a shear band (b) following homogeneous deformation (a) in the axisymmetric compression test.

Figure 2 shows schematically the formation of a compaction band. In Figure 2b, axial compression occurs uniformly (the distance between the horizontal lines in Figure 2a is decreased uniformly in Figure 2b). In Figure 2c, compression occurs predominantly in a narrow zone, i.e. a compaction band; outside this zone much less compression occurs. The plane of the compaction band is perpendicular to the largest compressive principal stress.

Antonellini & Aydin (1995) and Mollema & Antonellini (1996) have reported observations of compaction bands in porous sandstone formations. Olsson (1999) identified examples of compaction bands in earlier laboratory experiments and reported observations of compaction bands in axisymmetric deformation experiments on Castlegate sandstone. Some of Olsson's samples contained both compaction and shear bands. Other recent observations of compaction bands in experiments have been reported by Olsson & Holcomb (2000), DiGiovanni et al. (2000), Olsson (2001), Olsson et al. (2002), Haimson (2001), Klein et al. (2001) and Wong et al. (2001).

The appearance of the compaction bands is, however, much different in the experiments of different groups. Olsson & Holcomb (2000), Olsson (2001) and Olsson et al. (2002) have used acoustic emissions and permeability measurements to infer a widening zone of compaction (Fig.

3a). The zone begins near the ends of the specimen and spreads behind a planar front that propagates with a speed roughly an order of magnitude greater than the piston velocity. According to the analysis of Olsson (2001), the speed of the propagating front is proportional to the porosity difference between the compacted and uncompacted portions of the specimen. In contrast, Klein et al. (2001) and Wong et al. (2001) observe thin planar zones of compacted material alternating with uncompacted material (Fig. 3b). These zones also begin near the specimen ends; increasing piston displacement causes progressive formation of additional zones farther from the ends. The sandstones tested by the two groups are different and it is unclear whether this or some other aspect of the testing is causing the different appearances of the compaction zones.

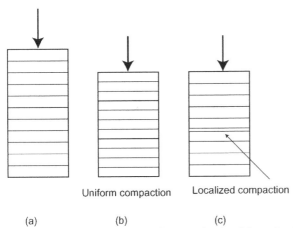

Uniform compaction   Localized compaction

(a)   (b)   (c)

Figure 2. Schematic illustration of compaction band formation in the axisymmetric compression test. (a) undeformed specimen; (b) uniform compaction; (c) localized compaction band formation.

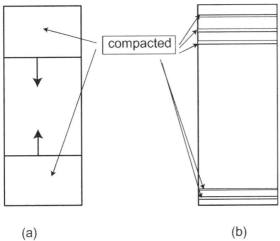

compacted

(a)   (b)

Figure 3. Different appearance of evolving compaction zones. In (a) the compacted zone widens and spreads across the specimen with increasing piston displacement; in (b) localized bands of compaction alternate with less compacted material; additional bands form farther from the ends as the piston displacement continues.

3 WHY DO COMPACTION BANDS FORM?

Olsson (1999) recognized that the inception of compaction bands could be addressed as a bifurcation from homogeneous deformation, the same approach used to study the inception of shear

31

bands (Rudnicki & Rice, 1975, Rice, 1976). For boundary conditions that permit homogeneous deformation as a possible solution for the next increment of deformation, this approach seeks conditions for which the formation of a band of localized deformation is an alternative solution. The formation of the band solutions is subject to conditions of compatibility and equilibrium.

The compatibility condition requires that velocities, or increments of displacement, are continuous at the instant of band formation. Consequently, the displacement increment within the band du_i^{band} is related to the increment in the homogeneous field outside the band du_i^0 by

$$du_i^{band} = du_i^0 + g_i(\mathbf{n}\ \mathbf{x})$$ (1)

where $\mathbf{n}$ is the unit normal to the plane of the band and the g_i are functions only of distance across the band. The corresponding strain increments are related by

$$d\varepsilon_{ij}^{band} = d\varepsilon_{ij}^0 + \frac{1}{2}(n_i g_j + g_i n_j)$$ (2)

For a compaction band the vector $\mathbf{g}$ is in the same direction as the normal but opposite in sense, $\mathbf{g} \propto -\mathbf{n}$. Thus, the difference in strain increments (2) is uniaxial.

Equilibrium requires that the increments of traction be continuous across the band interface at the onset of localization:

$$dt_i = n_j(d\sigma_{ji})^{band} = n_j(d\sigma_{ji})^0$$ (3)

If the constitutive relation can be expressed in the incrementally linear form

$$d\sigma_{ij} = L_{ijkl}d\varepsilon_{kl}$$ (4)

then substituting (4) into (3) and using (2) yields the condition for localization:

$$n_i L_{ijkl} n_l g_k = 0$$ (5)

In obtaining (5) it has been assumed, without loss of generality, that L_{ijkl} is symmetric with respect to interchange of the first two and last two indices, but not necessarily with respect to interchange of the first and last pair. The result (5) also assumes that the response at the onset of band formation is the same both inside and outside of the band but Rice & Rudnicki (1980) have shown that for a wide class of material models this bifurcation precedes one that occurs with elastic unloading outside the band (Chambon (1986) has given a more general derivation of this result). Rice (1976) has given a more general formulation of the localization problem.

A non-zero solution for the g_k (5), which expresses a nonuniform deformation field, is possible only if the determinant of the coefficient matrix vanishes. In general, the direction of $\mathbf{g}$ depends on the constitutive properties (Bésuelle, 2001), but for a compaction (or dilation) band, it is in the same direction as $\mathbf{n}$, i.e $\mathbf{g} = \alpha\mathbf{n}$ where α is a scalar. Substituting into (5) and multiplying by n_j reveals that the condition for inception of a compaction or dilation band is that the tangent modulus for uniaxial deformation relative to the plane of the band must vanish. Thus, for a compaction (or dilation) band (5) requires that

$$n_i n_j L_{ijkl} n_k n_l = 0$$ (6)

Olsson (1999) arrived at the same condition using a one-dimensional analysis and the results of Issen & Rudnicki (2000, 2001), who use the constitutive relation employed by Rudnicki & Rice (1975), can be expressed in the same form.

4 CONSTITUTIVE MODELS

4.1 Axisymmetric Compression

Whether the condition (6) or (5) can be met obviously depends strongly on the type of constitutive model, i.e. the form of the L_{ijkl}. To illustrate the dependence on constitutive parameters in a simple way, consider the relation used by Rudnicki (1977, 2002) for axisymmetric deformation. The standard axisymmetric compression test is shown schematically in Figure 4a. If, as in the most common test, the lateral stresses are held constant, the axial and lateral strain increments can be expressed as

$$d\varepsilon_{axial} = d\sigma_{axial}/E, \ d\varepsilon_{lat} = -vd\varepsilon_{axial} \tag{7}$$

Thus, E is the tangent modulus for constant lateral confining stress and v plays a role similar to Poisson's ratio but is a nonelastic parameter and may vary with the deformation (Fig. 4b). Additional constitutive parameters can be defined by increasing the lateral stress and adjusting the axial stress so that the axial deformation is held fixed. Then the increments of lateral strain and axial stress are expressed as

$$d\varepsilon_{lat} = 2d\sigma_{lat}/9K, \ d\sigma_{axial} = rd\sigma_{lat} \tag{8}$$

The ratios r and v are normally positive, but Figure 5 shows a simple, pin-jointed model structure for which r and v are negative. Although this might be a plausible microstructural model for a very porous, open-cell material, it would seem less appropriate for most rocks or soils. Nevertheless, it does make clear that negative r and v are possible. An expression for the tangent modulus for uniaxial strain (zero lateral strain) can be obtained by combining (7) and (8), $E_{uniaxial} = E + 9Krv/2$. Setting this value equal to zero yields the critical value of the tangent modulus for compaction band formation in a test with constant lateral confining stress:

$$E_{crit} = -9vrK/2 \tag{9}$$

Figure 6 shows a schematic illustration of a typical stress strain curve for a very porous sandstone (at relatively high confining pressure). Compaction bands, if they occur, form on the flat shelf of the stress strain curve where $E \approx 0$. Because the modulus K is presumably positive, (9) implies that either r or v is near zero at compaction band formation. Experimental evidence on this issue is mixed. Figure 6 of Olsson (1999) shows constant lateral deformation at the mid-height of an axisymmetric compression sample indicating $v \approx 0$ but Wong et al. (2001) report increasing lateral deformation indicating $v > 0$. In addition, values of the dilatancy factor, β of Rudnicki & Rice (1975) (see below), inferred by Olsson (1999) and Wong et al. (2001) correspond to increasing lateral deformation. The response in these experiments may, however, differ significantly from the isotropic response assume by the Rudnicki & Rice (1975) relation.

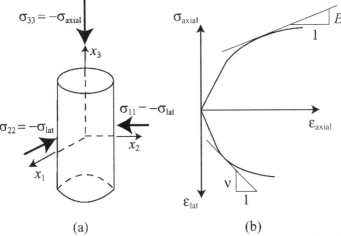

(a) (b)

Figure 4. Schematic illustration of the axisymmetric compression test (a) and definitions of the constitutive parameters E and v for constant lateral stress used by Rudnicki (2002).

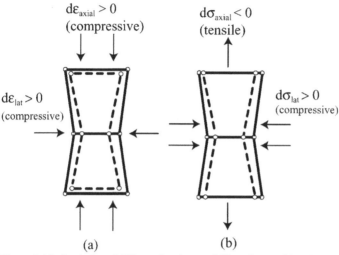

Figure 5. Mechanical model illustrating the possibility of $\mu < 0$ (a) and $r < 0$ (b).

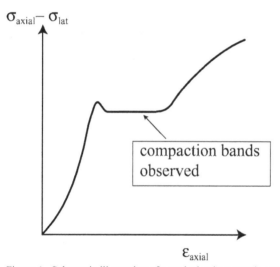

Figure 6. Schematic illustration of a typical axisymmetric compression stress – strain curve for a very porous sandstone.

4.2 General Deformation States

Analyses of the inception of compaction bands for general deformation states have been based on modifications and extensions of the analysis of shear band inception by Rudnicki & Rice (1975). They noted the possibility of a bifurcation corresponding to localized deformation in a planar zone perpendicular to one of the principal stresses (such as a compaction band), but they did not elaborate because this case occurred for a range of constitutive parameters outside those for the low porosity rocks on which they were focusing. Ottosen & Runesson (1991) also identified the possibility of this type of bifurcation for a slightly more general constitutive model than considered by Rudnicki & Rice (1975) and noted that it would occur for a material modeled with a yield surface "cap", which allows for yielding in hydrostatic compression (Dimaggio & Sandler, 1971). Ottosen & Runesson (1991) also found that the condition for the occurrence of this type of solution differed from that given by Rudnicki & Rice (1975). Perrin & Leblond (1993) corrected and discussed in detail this error in Rudnicki & Rice (1975).

Issen & Rudnicki (2000, 2001) give an extensive discussion of the possibilities for the occurrence of planar localized zones perpendicular to the most or least compressive principal stress and the interpretation of the former in terms of compaction zones. They do this within the constitutive framework used by Rudnicki & Rice (1975), the most general form for an incrementally linear elastic-plastic solid with yield surface and plastic potential depending on only the first and second stress invariants. Figure 7 shows such a yield surface described by

$$F(\tau,\sigma,H)=0 \tag{10}$$

where $\tau=\sqrt{s_{ij}s_{ij}/2}$ is the second invariant of the deviatoric stress, $s_{ij}=\sigma_{ij}-(1/3)\delta_{ij}\sigma_{kk}$, δ_{ij} is the Kronecker delta, $\sigma=-\sigma_{kk}/3$ is the mean normal stress (positive in compression) and H denotes a set of one or more variables that describe the current state of plastic deformation. Figure 7 also shows the inelastic increments of the volume strain $d\varepsilon^{p}$ (positive in extension) and equivalent shear strain $d\gamma^{p}$ as components of a vector that is normal to the plastic potential surface. The friction coefficient used by Rudnicki & Rice (1975), μ, is equal to the local slope of the yield surface $-(\partial F/\partial\sigma)/(\partial F/\partial\tau)$ and the dilatancy factor, β, is $d\varepsilon^{p}/d\gamma^{p}$. Because their primary interest was shear localization in low porosity, dilatant rocks, Rudnicki & Rice (1975) focused attention on the case β and μ positive. But, as shown in Figure 7 and as noted by Issen & Rudnicki (2000), and previously by Ottosen & Runesson (1991) using different notation, β and μ are negative on the "cap" portion of the yield surface.

The trajectory of the standard axisymmetric compression test in Figure 7 is a straight line with slope $\sqrt{3}$ that intersects the horizontal axis at the value of the constant lateral confining stress (Fig. 8). Consequently, tests at low confining stress will intersect the shear surface first and tests at higher confining stresses will intersect the cap first.

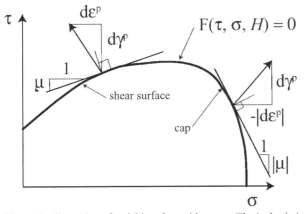

Figure 7. Illustration of a yield surface with a cap. The inelastic increment of strain is plotted as a vector. This vector is normal to the plastic potential surface, but not, in general, to the yield surface.

Issen & Rudnicki (2000), incorporating the correction of Perrin & Leblond (1993) find that compaction bands will precede shear bands if

$$\frac{1}{3}(\beta+\mu)\leq\frac{(1-2\nu)N-\sqrt{4-3N^{2}}}{2(1+\nu)} \tag{11}$$

where ν is Poisson's ratio (equal to ν in (7) and (9) only for elastic deformation) and N is a deviatoric stress state parameter equal to the intermediate principal deviatoric stress divided by 2τ. (An equivalent condition is given by Ottosen & Runesson (1991)). For axisymmetric compression $N=1/\sqrt{3}$ and (11) reduces to

$$\beta+\mu\leq-\sqrt{3} \tag{12}$$

Thus, compaction band formation is favored on the "cap" portion of the yield surface where β and μ are negative. The corresponding critical value of the tangent modulus for constant lateral confining stress (obtained by combining (19) and (21) of Issen & Rudnicki (2000)) is

$$\frac{E}{2G(1+\nu)} = \frac{-(1+2\mu/\sqrt{3})(1+2\beta/\sqrt{3})}{(1-\nu)(1-\mu/\sqrt{3})(1-\beta/\sqrt{3})-(1+2\mu/\sqrt{3})(1+2\beta/\sqrt{3})} \tag{13}$$

where G is the elastic shear modulus and the combination in the denominator of the left hand side is Young's modulus.

The cap portion of the yield surface in Figure 7 connects smoothly to the shear surface. However, there is evidence (e.g. Wong et al. 1997, Wong & Baud, 1999) that the shear and cap portions are separate surfaces and they are often implemented as such in numerical calculations (e.g., Fossum et al. 1995, Fossum & Frederich, 2000). The simplest interpretation of the two surfaces is that they reflect different microscale mechanisms of inelasticity: shear-driven local tensile cracking and dilatancy for the shear surface and pore collapse and grain crushing for the cap. The evolutions of the cap and shear surfaces will be different, in general, and depend on different measures of inelastic deformation (e.g. accumulated inelastic volume strain or accumulated inelastic equivalent shear strain). Wong et al. (2001) have interpreted their data on Bentheim sandstone as requiring activation of multiple damage mechanisms. Figure 8 illustrates a yield surface in which the cap and shear portions intersect at a corner.

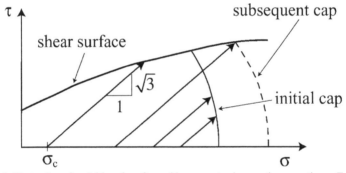

Figure 8. Illustration of a yield surface formed by separate shear and cap portions. Four trajectories of axisymmetric compression tests with different constant lateral confining pressure (σ_c) are also shown.

The additional flexibility of the direction of the inelastic strain increment vector at a yield surface corner is well known to enhance the possibility of localization (Rudnicki & Rice, 1975). Although it is unlikely that the stress path will intersect the corner on the initial yield surface, if the two surfaces evolve separately, increasing loads will cause the stress state to coincide with the corner (as illustrated by one trajectory in Fig. 8). The analysis of bifurcation conditions at a corner are much more complicated but some limiting results have been given by Issen (2002).

5 CONCLUDING DISCUSSION

Compaction bands have been observed in porous sandstones in the field and in laboratory axisymmetric compression tests on several varieties of sandstones. A bifurcation approach demonstrates that compaction bands are a possible alternative solution to homogeneous deformation for constitutive parameters that are appropriate for high porosity rock. In particular, inelastic volume compression and a shear yield stress that decreases with increasing mean compressive stress favor formation of compaction bands. These features are characteristic of a "cap" yield surface often used to model the inelastic response of porous geomaterials.

There are, as yet, many open questions about the formation and evolution of compaction bands. Although the bands observed in the laboratory are similar to those observed in the field, their length scale is limited to the size of the laboratory specimen and, therefore, is much smaller. Furthermore, the bands observed in the laboratory tend to form first near the specimen ends and, consequently, appear to be triggered by slight nonuniformities caused by the end constraint. For these reasons, there remain questions about whether band formation in the field and laboratory are the same phenomenon.

Predictions for the inception of compaction band formation from bifurcation theory agree roughly with observations in the following sense: compaction band formation is predicted for constitutive parameters representative of porous sandstones. Numerical values of these parameters inferred from experiments are close to but do not agree in detail with those predicted by analysis. This discrepancy may be due to inadequacy of the constitutive model, inappropriateness of the bifurcation approach or uncertainties in the numerical determination of parameters. Predictions are sensitive to details of the structure and evolution (with nonelastic deformation) of the cap surface and the nature of its intersection with the shear surface. These features are not, at present, well-constrained by experiments.

Compaction band formation clearly requires high porosity (sandstones in which they have been observed range from about 15 to 30% porosity). There is, however, no theory explicitly connecting predictions to porosity, nor is it known from either experiments or theory whether there exists some critical minimum value of porosity. Because the porosity of natural materials cannot be controlled and is limited in range, this issue might be addressed profitably by experiments on man-made sintered materials. Although compaction band formation is roughly correlated with phenomenological nonelastic parameters, e.g. those of the cap model, little understanding exists of how band formation depends on the microstructure, e.g. grain size, shape, and distribution, cement and presence of clay minerals. Vardoulakis and Sulem (1995) do suggest, however, that compaction instabilities are more prevalent in geomaterials of uniform grain and pore sizes. Furthermore, DiGiovanni et al. (2000) have noted some differences in the micromechanical deformation mechanisms of Castlegate sandstone, in which compaction bands have been observed, from those of Berea sandstone (Menéndez et al. 1996)in which they have not.

The bifurcation analysis addresses the inception of band formation but not the subsequent evolution. Experiments have revealed that the evolution can have at least two quite different structures: widening of the compacted zone and widening of a pattern of compacted and uncompacted regions. Reasons for these different morphologies are not known.

The bifurcation analysis also suggests that there should be a transition from compaction bands to shear bands at high angles to the specimen axis for axisymmetric compression tests with decreasing values of the lateral confining pressure. Although both compaction bands and shear bands have been observed in some experiments, observations of very high angle shear bands are infrequent. More generally the nature of the transition between shear and compaction bands is unclear and likely depends on the nature of the intersection of the shear and cap surfaces.

The primary practical significance of compaction band formation and evolution is their role in altering the permeability structure of a formation. Experiments (Papamichos et al., 1993, Holcomb & Olsson, 2002) have demonstrated that compaction significantly reduces the permeability, but otherwise the effects of interaction of pore fluid have not been explored. In a different context, theoretical study of compaction of a fluid-saturated material with a nonlinear dependence of permeability on porosity suggests the possibility of solitary wave solutions (Barcilon & Richter, 1986). Predictive capability for field applications requires a better understanding of the role of pore pressure effects on evolution of the cap surface and of nonlinear effective stress effects on the permeability of compacting material.

Acknowledgement. I am grateful for numerous discussions with David Holcomb, Bill Olsson and Teng-fong Wong. I am also grateful to Lallit Anand for bringing to my attention the connection with non-uniform compaction of sintered materials. Partial financial support for this research has been provided by the U. S. Department of Energy, Office of Basic Energy Sciences under Grant DE-FG-02-93ER14344 to Northwestern University.

REFERENCES

Antonellini, M. & Aydin, A. 1995. Effect of faulting on fluid flow in porous sandstones: geometry and spatial distribution. *Am. Assoc. Petrol. Geol.* 79: 642-671.

Barcilon, V. & Richter, F. M. 1986. Nonlinear waves in compacting media. *J. Fluid Mech.* 164: 429-448.

Bastawros, A-F., Bart-Smith, H. & Evans, A. G. 2000. Experimental analysis of deformation mechanisms in a closed-cell aluminum alloy foam. *J. Mech. Phys. Solids* 48:301-322.

Bésuelle, P. 2001. Compacting and dilating shear bands in porous rock: theoretical and experimental conditions. *J. Geophys. Res.* 106: 13435-13442.

Chambon, R. 1986. Bifurcation par localization en bande de disaillement, une approache aved des lois incrémentalement non linéaires. *J. Méc. Théor. Appl.* 5(2): 277-298.

DiGiovanni, A. A., Fredrich, J. T., Holcomb, D. J. & Olsson, W. A. 2000. Micromechanics of compaction in an analogue reservoir sandstone. In J. Girard, M. Liebman, C. Breeds, & T. Doe (eds.), *Pacific Rocks* 2000: 1153-1158. Rotterdam: Balkema.

Dimaggio, F. L. & Sandler, I. S. 1971. Material model for granular soils. *J. Engn. Mech., ASCE* 97: 935-950.

Fossum, A. F., Senseny, P. E., Pfeifle, T. W. & Mellegard, K. D. 1995. Experimental determination of probability distributions for parameters of a salem limestone cap plasticity model. *Mech. Mat.* 21:119-137.

Fossum, A. F. & Fredrich, J. T. 2000. Cap plasticity models and compactive and dilatant pre-failure deformations. In J. Girard, M. Liebman, C. Breeds, & T. Doe (eds.), *Pacific Rocks* 2000: 1169-1176. Rotterdam: Balkema.

Gu, C., Kim, M. & Anand, L. 2001. Constitutive equations for metal powders: application to powder forming processes. *Int. J. Plasticity* 17: 147-209.

Haimson, B. C. 2001. Fracture-like borehole breakouts in high-porosity sandstone: are they caused by compaction bands? *Phys. Chem. Earth (A)* 26: 15-20.

Holcomb, D. J. & Olsson, W. A. 2002. Compaction localization and fluid flow. Submitted to *J. Geophys. Res.*

Issen, K. A. & Rudnicki, J. W. 2000. Conditions for compaction bands in porous rock. *J. Geophys. Res.* 105: 21529-21536.

Issen, K. A. & Rudnicki, J. W. 2001. Theory of compaction bands in porous rock. *Phys. Chem. Earth (A)* 26:95-100.

Issen, K. A.. 2002. The influence of constitutive models on localization conditions for porous rock, *Engn. Frac. Mech.*, to appear.

Klein, E., Baud, P., Reuschle, T., & Wong, T-f. 2001. Mechanical behaviour and failure mode of Bentheim sandstone under triaxial compression. *Phys. Chem. Earth (A)* 26: 21-25.

Menéndez, B., Zhu, W., & Wong T.-f. 1996. Micromechanics of brittle faulting and cataclastic flw in Berea sandstone. *J. Struct. Geol.* 18(1): 1-16.

Mollema, P. N. & Antonellini, M. A. 1996. Compaction bands: a structural analog for anti-mode I cracks in aeolian sandstone. *Tectonophysics* 267: 209-228.

Nagel, N. B. 2001. Compaction and subsidence issues within the petroleum industry: from Wilmington to Ekofisk and beyond. *Phys. Chem. Earth (A)* 26: 3-14.

Olsson, W. A. 1999. Theoretical and experimental investigation of compaction bands in porous rock, *J. Geophys. Res.* 104: 7219-7228.

Olsson, W. A. 2001. Quasistatic propagation of compaction fronts in porous rock. *Mech. Mat.* 33: 659-668.

Olsson, W. A. & Holcomb, D. J. 2000. Compaction localization in porous rock. *Geophys. Res. Letters,* 27: 3537-3540.

Olsson, W. A., Holcomb, D. J. & Rudnicki, J. W. 2002. Compaction localization in porous sandstone: implications for reservoir compaction. *Oil & Gas Science and Technology, Revue de l'institut Français du Pétrole* 57 (5): 591-599.

Ottosen, N. S. & K. Runesson, K. 1991. Properties of discontinuous bifurcation solutions in elastoplasticity. *Int. J. Solids Struct.* 27 (4): 401-421.

Papamichos, E., Vardoulakis, I. & Quadfel, H. 1993. Permeability reduction due to grain crushing around a perforation. *Int. J. Rock Mech. Min. Sci. Geomech. Abstr.* 30:1223-1229.

Papka, S. D. & Kyriakides, S. 1998. In-plane crushing of a polycarbonate honeycomb. *Int. J. Solids Struct.* 35: 239-267.

Park, C. & Nutt, S. R. 2001. Anisotropy and strain localization in steel foam. *Mat. Sci. Eng.* A299: 68-74.

Perrin, G. & LeBlond, J. B. 1993. Rudnicki and Rice's analysis of strain localization revisited. *J. Appl. Mech.* 60: 842-846.

Rice, J. R. 1976. The localization of plastic deformation. In W. Koiter (ed.), *Theoretical and Applied Mechanics, Proc. 14th IUTAM Congress*: 207-220. Amsterdam: North Holland.

Rice, J. R. & Rudnicki, J. W. 1980. A note on some features of the theory of localization of deformation. *Int. J. Solids Struct.* 16: 597-605.

Rudnicki, J. W. 1977. The effect of stress-induced anisotropy on a model of brittle rock failure as localization of deformation. In F.-D. Wang & G. B. Clark (eds.), *Energy Resources and Excavation Technology, Proceedings of the 18th U. S. Symposium on Rock Mechanics*: 3B4- - 3B4-8.

Rudnicki, J. W. 2002. Conditions for compaction and shear bands in a transversely isotropic material. *Int. J. Solids Struct.* 39 (13-14): 3741-3756.

Rudnicki, J. W. & Rice, J. R. 1975. Conditions for the localization of deformation in pressure-sensitive dilatant materials. *J. Mech. Phys. Solids* 23: 371-394.

Vardoulakis, I. & Sulem, J. 1995. *Bifurcation Analysis in Geomechanics*. Glasgow: Blackie Academic.

Wawersik, W. R., Rudnicki, J. W., Dove, P., Harris, J., Logan, J. M., Pyrak-Nolte, L., Orr, F. M., Jr., Ortoleva, P. J., Richter, F., Warpinski, N., Wilson, J. L. & Wong, T.-f. 2001. Terrestrial sequestration of CO2: An assessment of research needs. In *Advances in Geophysics* 43: 97-177.

Wong, T-f., Baud, P. & Klein, E. 2001. Localized failure modes in a compactant porous rock, *Geophys. Res. Lett.* 28: 2521-2524.

Bifurcations & Instabilities in Geomechanics, – Labuz & Drescher (eds.)
© 2003 Swets & Zeitlinger, Lisse, ISBN 90 5809 563 0

Further remarks on well-posedness within hypoplasticity for granular materials

Zbigniew Sikora & Rafal Ossowski
Geotechnical Department, Hydro- & Environmental Eng. Faculty
Technical University of Gdansk, Poland

ABSTRACT

Correct formulation of the differential equation system for equilibrium conditions in granular materials especially of controlled numerical calculation view-point will be discussed. Rather simple algorithm for necessary conditions of well-posedness is proposed. Suggestion for implementation into a computer program is also proposed. It is known from the theory of partial differential equations, that the solution of a differential equation system is not uniquely determined by the equation system itself, so therefore one has properly to formulate initial and boundary conditions, which completely determine a solution of the differential equation system. However, apart from this there is still another supplementary consideration to be worked out, which plays an important role especially in numerics. If we consider that a certain amount of error in the initial or boundary conditions is inescapable, then this error will manifest itself in the solution too. This problem is not trivial because it will not always follow that the error in the solution will also be small. The last two sentences treat, broadly speaking, the dependence of the solution on the boundary conditions or, in other words, the stability of the solution. In order to formulate initial and boundary problems three conditions must be checked, namely existence, uniqueness and stability of the solution. The set of the three conditions guarantees that the differential problem is correctly formulated or the problem is well-posed. If we look into the question of solution stability in an unbounded domain, we find that even for ordinary differential equations the question ceases to be simple. The related questions in the theory of partial differential equations could also fittingly be called the problems of stability. So far these questions have not been fully answered; they are too difficult for an elementary study. The object of this paper is to derive some conditions, which assert compliance with restrictions of well-posedness following a theory proposed by the first author.

1 SHORTCOMINGS IN HYPOPLASTICITY FOR SOILS

Hypoplasticity for soils is an incrementally nonlinear rate-type constitutive model of granular materials. It is able to describe the dissipative behavior, the plastic flow and nonlinear effects within the yield surface with a single tensorial equation. Early versions of the hypoplastic constitutive model, in spite of their simplicity, performed quite well in many numerical applications including oedometric or triaxial element tests. Their great advantage over the elastoplastic formulation was the „perfect" nonlinearity which facilitated spontaneous localization of deformation. However, the first hypoplastic models allowed high level stress ratios during some loading programs (lack of bound surface), they generated high pore pressure during undrained cyclic shearing and allowed for an enormous accumulation of strain during small stress cycles, which is a result of ratchetting effect. The stiffness was a function of stress, i.e. path-independency. The influence of density, so called pyknotropy, was not considered and the assumption of proportional increase of stiffness with the mean stress was simplified.

These shortcomings manifested themselves in advanced applications of the model, however a lot of shortcomings have been already improved, although, e.g. predictions of deformation involving creep effects were inaccurate, which is of importance for constructions susceptible to settlements. The performance of the hypoplastic constitutive model was not satisfactory in the FE-calculation with cyclic loading. In order to overcome these and similar problems several extensions to hypoplasticity have been proposed.

A geotechnical engineer may ask whether a sophisticated constitutive model is necessary for his practical decisions and what degree of complexity is reasonable for the problem faced. A systematic decisions of practical needs in geotechnical engineering was recently presented by Simpson [10]. Although in many cases both the recommendation of practice and the constitutive model may be based on the same empirical knowledge, the numerical model expresses this knowledge in a precise mechanical and mathematical language, so one can be optimistic about the further applications of highly-developed constitutive equations. Due to rapid development of computational systems and capacities of existing nowadays finite element programs, the perspectives of solving even complex geotechnical boundary value problems are very promising.

However there is still lack of effective numerical procedures, which can guarantee that from mechanical and mathematical point of view the solution of an initial-boundary value problem is correct. One of the important problems in the framework of the partial differential equation theory which order to preserve type of the governing equation during the numerical evaluation of the initial value problem. There exists many stabilized finite element formulations, for single or multiple-phase geomaterials, or many regularization methods, which as a matter of fact improve the FE calculations but the problem of changing the type of the governing system remains still open in numerical monitoring of the evaluation process.

Available auxiliary FE-codes, e.g. GID (*E. Oñate, CIMNE*) being a pre- & postprocessor, make today easier the FE-analysis with own codes. Many young researchers construct their own computational procedures based on FEM or other numerical integration procedures, since the basic numerical knowledge in the framework is mostly requirement of the technical PhD-courses today. In this context the proper mathematics of a numerical solution is of importance. However many mathematical requirements can not be checked on an explicit one, therefore some procedures for checking for correctness of the numerical solution are the essential ones.

Effective computer codes with procedures dedicated to soil mechanics are widely available, e.g. in the commercial field of finite element programs ANSYS, PATRAN, ABAQUS, MARC, DIANA, PLAXIS, TOCHNOG or INCNL2D[1]. Most of them accept a constitutive description provided by the user, which is a good prospect for the development of hypoplasticity.

---

[1]Author's FEM code: **INC**rementally **N**on **L**inearity in **2D**

42

Summing up: the FE-results can clearly support the conventional design methods but the quality of the constitutive theory is of cruicial importance in this context. Thus, procedures on well-posedness conditions, which make the numerical analysis reasonable in mechanical viewpoint, should or even must be still improved due to the new versions of hypoplasticity, in particular.

2 CONDITIONS FOR A WELL-POSED BVP

Several hypoplastic models and their extensions can be found in the literature. Most of the proposed concepts were succesfully implemented into FE codes.

However there is some difficulties in applying elastoplasticity to geomaterials, e.g. decomposition of deformation into elastic and plastic parts and the transition between elastic and plastic deformation. An unfortunate formal decomposition of the stress rate into linear and nonlinear part, commonly used by the hypoplastic community makes the hypoplastic model difficult to understand, especially for those accustomed to the elastoplastic terminology. Hypoplasticity for granular materials has been proposed originally by Kolymbas, s. e.g. [7]. Some concepts pertinent to elastoplasticity, e.g. yield surface, plastic potential, decomposition of the deformation into elastic and plastic parts, were abandoned to be used in formating the constitutive equations. The theoretical framework of hypoplastic theory is outlined following the recent work by Kolymbas, [7], and others [9, 13, 14].

Nevertheless, some realistic problems have been already solved based on correct defined boundary value problems within hypoplasticity, [9].

There is a necessity that a BVP should be well-posed, in particular from physical and also numerical viewpoint. From theoretical viewpoint three conditions due to the solution must be fulfilled: **existance, uniqueness** and **stability**. If we take into consideration Sobolew's statement, namely that „.... *the solution of problems, which are ill-posed has no practical value in the majority of cases*", [11, 12], then the problem of uniqueness and stability becomes as very important one.

Each problem in application contains several physical parameters which may vary over a certain specified set. Thus, it is important to understand the qualitative behaviour of the system as the parameters vary. A good design for a system will always be such that the qualitative behaviour does not change when the parameters are varied a small amount the value for which the original design was made. However, the behaviour may change when the system is subjected to large variations in the parameters. A change in the qualitative properties could mean a change in stability of the original system and thus the system must assume a state different from the original design. The value of the parameters where this change takes place are called **bifurcation values** or **bifurcation points**. Knowledge of the bifurcation points is therefore absolutely necessary for the complete understanding of the system[2]. Our objective is to present a method to check whether the system reaches a bifurcation point.

---

[2]„The bifurcation theory is a key to understanding the complex behaviors in the nature", N.N. Moisiejev "

3 A SHORT CLASSI CATION OF BIFURCATION MODES

Depending on the physical complexity of the system, the stationary system with extremely small increments of loading parameter λ (active parameter) may come at the bifurcation point

- **smoothly** on the post-bifurcation branch-path, which has been chosen by the imperfections. Such bifurcation is called **divergent bifurcation** or **safe bifurcation**. The critical value of the loading parameter λ is to be calculated from the appropriate eigenvalue problem of the system; or

- **dynamicaly** to the post-bifurcated mode, which is of oscillatory character. Such bifurcations are called **Hopf bifurcations** or **flatters** in solid mechanics. In these cases the bifurcation is characterized by two quantities: critical quantity of active parameter (e.g. Reynolds number, or Mach number) and oscillation frequency of post-bifurcated mode.

In general, two distinct aspects of bifurcation theory can be discussed – static and dynamic.

Static bifurcation theory is concerned with the changes that occur in the structure of the set of zeros of a function as parameters in the function are varied. If the function is a gradient, then variational techniques play an important role and can be employed even for global problems. If the function is not a gradient or if more detailed information is desired, the general theory is usually local. At the same time, the theory is constructive and valid when several independent parameters appear in the function. In differential equations, the equilibrium solutions are the zeros of the vector field. Therefore, methods in static bifurcation theory are directly applicable.

Dynamic bifurcation theory is concerned with the changes that occur in the structure of the limit sets of solutions of differential equations as parameters in the vector field are varied. For example, in addition to discuss the way that the set of zeros of the vector field (the equilibrium solutions) changes through the static theory, the stability properties of these solutions must be considered. In fact, there is an intimate relationship between changes of stability and bifurcation. The dynamics in a differential equation can also introduce other types of bifurcations: e.g. periodic orbits, homoclinic orbits, invariant tori. This fact introduces several difficulties which may require rather advanced topics in differential equation theory.

4 USEFUL TOOLS IN BIFURCATION ANALYSIS & LINEARIZATION TECHNIQUE

A sufficient condition for a bifurcation point of an initial value problem with a hypoplastic constitutive equation, based on Hadamard theory of wave propagation, can be studied e.g. in [9]. However, the bifurcation condition, defines tensors of stress and streching respectively, which may posibile the onset of a shear band. Such bifurcation condition plays an important role in numerical viewpoint, in particular. These conditions can be checked at least at each Gauss point of the discretized volume and can be checked at each time reference during the evolution of the initial-boundary value problem. However, the mentioned condition posesses a disadventage, which consists of a description of the so called *jump parameter*. This parameter depends on the solution, so the bifurcation process could be discussed only as a non-trivial nonlinear optimization problem.

In bifurcation analysis of initial-boundary value problems constitutive equation plays the crusial role. In this context, it is of considerable interest for a given constitutive equation to check whether bifurcation of the solution is possible and in which direction will be continued. These questions are not trivial, so there has been ample research concerning stability criteria in the mathematical and mechanical literature, [2, 3, 1, 6] and many others. However, a little has been already done to apply these strong mathematical criteria to a specific family of constitutive equations. Owing to the great complexity of the most constitutive equations, the bifurcation condition can be hardly applied analytically. In the following we utilize some known theorems on bifurcation and present the possibility to use them in the numerical calculations.

4.1 *The hypoplastic equations*

Consider the following problem:

$$\dot{\mathbf{T}} = \mathbf{h}(\mathbf{T}, \mathbf{D})$$
$$\mathbf{T}(t_0) = \mathbf{T}_{t_0} \tag{1}$$

In fact under $\mathbf{h}$ one can indicate a constitutive equation, formulated on the representation theorem of an isotropic tensor-valued function. The general form of a constitutive equation within hypoplasticity can be written as follows:

$$\dot{\mathbf{T}} = \overset{\circ}{\mathbf{T}} + \mathbf{WT} - \mathbf{TW} \tag{2}$$
$$\overset{\circ}{\mathbf{T}} = \mathbf{h}(\mathbf{T}, \mathbf{D}) = \mathbf{L}(\mathbf{T}, \mathbf{D}) + \mathbf{N}(\mathbf{T})\|\mathbf{D}\| \tag{3}$$

or proposed by Gudehus-Bauer, [4], as

$$\overset{\circ}{\mathbf{T}} = \mathbf{H}(e, \mathbf{T}, \mathbf{D}) \tag{4}$$
$$\overset{\circ}{\mathbf{T}} = a_1 f_b f_e \left[\mathbf{L}(\mathbf{T}, \mathbf{D}) + f_d f_a \mathbf{N}(\mathbf{T}, \mathbf{D}) \right] \tag{5}$$

where a_1 and f_a, f_b, f_e, f_d are material factors standing for argotropy, barotropy and pyknotropy respectively. $\mathbf{T}, \mathbf{D}$ and $\mathbf{W}$ are Cauchy stress tensor, rate of deformation (stretching) and spin tensor respectively. e stands for void ratio. The symbols $(\dot{\cdot})$ and $(\overset{\circ}{\cdot})$ indicate time derivation and Jaumann time derivative respectively.

In order to describe irreversible deformation, equation (3) should be nonlinear in relation to $\mathbf{D}$. The equation (3) shows this fact supposing that the $\mathbf{h}$ function is a sum of two tensor-valued functions $\mathbf{L}$ and $\mathbf{N}$. $\mathbf{L}$ denotes bilinear tensor-valued operator and $\mathbf{N}$ is a tensor-valued function of the stress tensor $\mathbf{T}$. $\| \cdot \|$ stands for the Euclidean norm of the stretching tensor. Since the functions $\mathbf{L}$ and $\mathbf{N}$ are isotropic ones then their specific forms can be obtained by invoking the representation theorem of an isotropic tensor–valued function.

Remark 1 *Because of* $\|\mathbf{D}\|$*, the constitutive equation (3) is incrementally nonlinear in* $\mathbf{D}$ *and therefore equation (3) defines a class of incrementally nonlinear constitutive equations.*

Let consider a complementary, locally parameterized family differential equations to (1), which consists of sum of the linear part, $\mathbf{A}(\mathbf{D})\mathbf{T}$, and the polynomial one, $\phi(\mathbf{T}, \mathbf{D})$, which has the order greater then one in relation to $\mathbf{T}$, i.e.

$$\dot{\mathbf{T}} = \mathbf{h}(\mathbf{T}, \mathbf{D}) \cong \bar{\mathbf{h}}(\mathbf{T}, \mathbf{D}) = \mathbf{A}(\mathbf{D})\mathbf{T} + \phi(\mathbf{T}, \mathbf{D}) \tag{6}$$

where for simplicity

$$\begin{array}{l} \bar{\mathbf{h}} : R^3 \times R^3 \to R^3 \\ \bar{\mathbf{h}}(\mathbf{0}, \mathbf{D}) = \mathbf{0} \end{array} \tag{7}$$

In particular, the fixed-point $\mathbf{T}_0$, for which $\mathbf{h}(\mathbf{T}_0, \mathbf{D}) = \mathbf{0}$ is of importance because with this point constant solutions $\mathbf{T}(t) \equiv \mathbf{T}_0$ are joined. Stability of the fixed-point $\mathbf{T}_0$ can be analyzed through linearization technique.

4.2 Linearization technique

We get linear differential equation in the neighbourhood of the point $\mathbf{T}_0$ using the following transformation:

$$\bar{\mathbf{T}}(t) = \mathbf{T}(t) - \mathbf{T}_0 \tag{8}$$

where the small imperfections can be analysed on the following equation

$$\dot{\bar{\mathbf{T}}} = \dot{\mathbf{T}} = \mathbf{h}(\mathbf{T}_0 + \mathbf{T}, \mathbf{D}) = \mathbf{h}'(\mathbf{T}_0)\bar{\mathbf{T}} + \dots \tag{9}$$

i.e.

$$\dot{\bar{\mathbf{T}}} = \mathbf{A} \cdot \bar{\mathbf{T}} + o(\|\bar{\mathbf{T}}\|); \qquad \mathbf{h}'(\mathbf{T}_0) \equiv \mathbf{A} := \left(\frac{\partial h_i}{\partial T_k}(\mathbf{T}_0) \right)^n_{i,k=1} \tag{10}$$

with matrix $\mathbf{A}$ as a Fréchet derivative of the tensor(vector)-valued function $\mathbf{h}$ at stress point $\mathbf{T}_0$. The main idea of (10) is that the summands of higher order $o(\|\bar{\mathbf{T}}\|)$ in the neighbourhood of $\bar{\mathbf{T}}_0 = \mathbf{0}$ are negligible and do not make any influence on the topological structure of the solution path. Whether it is true or not one needs a long mathematical proof — this fact is often supposed as an evident one on the intuitive way.

Thus we analyse a linear differential equation with constant coefficients of matrix $\mathbf{A}$

$$\begin{array}{rl} \dot{\bar{\mathbf{T}}} & = \mathbf{A} \cdot \bar{\mathbf{T}} \\ \bar{\mathbf{T}}(t_0) & = \bar{\mathbf{T}}_{t_0} \end{array} \tag{11}\tag{12}$$

The above problem (11-12) can be formulated at each Gauss point. There is no limitations if we consider the problem in the main stress configuration. Therefore the following two theorems, [5], can be useful for an analysis of solution's stability.

Theorem 1 *Let $\lambda_1, \lambda_2, \lambda_3$ are the eingenvalues of the matrix $\mathbf{A}$ of the linear equation (11),*

1. *if $\Re(\lambda_k) < 0 \quad \forall_{k=1,\ldots,3}$ then the fixed-point $\mathbf{T}_0$ of the main nonlinear problem (1) is asymptotically stable,*

2. *if $\Re(\lambda_i) > 0$ for one $i \in \{1, \ldots, 3\}$ then the fixed-point $\mathbf{T}_0$ is unstable, see also Appendix.*

Remark 2 *If there is one or more real parts equal to zero, then one requires to take into consideration the next higher order of the nonlinear summand of the expansion.*

The above theorem plays an important role and is an important tool in the qualitative discussion about solution's structure of the differential equation system! Its expression, that the linear equation provides a sufficient information about the trajectory pattern in the neighbourhood of $\mathbf{T}_0$, can be precisely stated in the next theorem.

Theorem 2 *If the matrix $\mathbf{A}$ of the linearized differential equation (11) do not have any eigenvalues with vanishing real part, then there exists homeomorphism[3] on the nieghbourhood $U(\mathbf{T}_0)$ of the point $\mathbf{T}_0$, which maps the trajectories of the non-linear pattern of $\mathbf{T}(t)$ on the linear one $(e^{t\mathbf{A}})$.*

If one is going to use the above theorems in numerical analysis, first of all the following characteristic equation, i.e.

$$\det(\mathbf{A} - \lambda \mathbf{I}) = 0 \tag{13}$$

must be solved. The equation (13) can be rewritten into the following polynomial equation

$$\sum_{k=0}^{n} a_{n-k} \lambda^k = 0; \qquad a_0 \doteq 1 \tag{14}$$

with real coefficients $a_1, \ldots, a_n$. Generally this equation can not be solved on an explicit way. In order to check whether all roots of the equation have the negative real parts one can use many algebraic criteria. Here we formulate the so-called Hurwitz criterion, which is very efficient in numerics.

Theorem 3 (Hurwitz criterion) *The eigenvalues $\lambda_1, \ldots, \lambda_n$ of the matrix $\mathbf{A}$ have all together negative real parts, if for the coefficients $a_1, \ldots, a_n$, from the polynomial (14), the following inequalities hold (let $a_k \equiv 0$ for $k > n$):*

$$D_1 \doteq a_1 > 0, D_2 \doteq \begin{Vmatrix} a_1 & 1 \\ a_3 & a_2 \end{Vmatrix} > 0, \ldots, D_n \doteq \begin{Vmatrix} a_1 & 1 & 0 & . & . & . & 0 \\ a_3 & a_2 & a_1 & 1 & 0 & . & 0 \\ . & & & . & . & . & . \\ a_{1n-1} & a_{2n-2} & . & . & . & . & a_n \end{Vmatrix} > 0 \tag{15}$$

Remark 3 *Since $D_n = D_{n-1} a_n$ the last inequality can be replaces by $a_n > 0$.*

Remark 4 *The easiness of the Hurwitz criterion pays its price! Since one does not know whether the eigenvalues are real or complex, single or multiple, therefore a precise discussion of the trajectory pattern needs more numerical knowledge. However, this problem will not be presented here.*

---

[3]Homeomorphism is an one-to-one mapping with its continuous inversion.

5 GOVERNING EQUATIONS AND THEIR LINEARIZATION

For simplicity we discuss problem of one-phase medium, a similar analysis is possible for two-phase medium as well.

If we look at a numerical element test the strains and stresses are homogeneous throughout the sample, which from the mathematical viewpoint corresponds to the solution of an equilibrium equation system in which the strain rate $\mathbf{D}$ and the stress $\mathbf{T}$ are functions of time alone and they are independent of the spatial coordinates.

We assume that the set of functions:

$$\mathbf{v}_0 = \mathbf{v}(t_0, \mathbf{x}), \mathbf{D}_0 = \frac{1}{2}(\nabla \mathbf{v}_0 + (\nabla \mathbf{v}_0)^T) \quad \text{and} \quad \mathbf{T}_0 \tag{16}$$

represents a homogeneous solution of the nonlinear equilibrium equation system

$$\begin{aligned} \rho \frac{\partial \mathbf{v}}{\partial t} - \operatorname{div}(\mathbf{T}) &= 0 \\ \frac{\partial \mathbf{T}}{\partial t} - \mathbf{h}(\mathbf{T}, \mathbf{D}) &= 0 \end{aligned} \tag{17}$$

We linearize the equation system (17) in the neighbourhood of (16) and obtain:

$$\mathbf{v}(t, \mathbf{x}) = \mathbf{v}_0 + \bar{\mathbf{v}}(t, \mathbf{x}) + \mathbf{R}_v$$

$$\mathbf{T}(t, \mathbf{x}) = \mathbf{T}_0 + \bar{\mathbf{T}}(t, \mathbf{x}) + \mathbf{R}_T \tag{18}$$

$$\rho \frac{\partial(\mathbf{v}_0 + \bar{\mathbf{v}})}{\partial t} - \operatorname{div}(\mathbf{T}_0 + \bar{\mathbf{T}}) = 0$$

$$\frac{\partial \mathbf{T}_0}{\partial t} + \frac{\partial \bar{\mathbf{T}}}{\partial t} = \mathbf{h}(\mathbf{T}_0 + \bar{\mathbf{T}}, \mathbf{D}_0 + \bar{\mathbf{D}}) \tag{19}$$

Suppose, as previously, that $\mathbf{v}_0$ and $\mathbf{T}_0$ are the homogeneous solutions. Thus taking into account that $\mathbf{h}$ is a homogenous constitutive function, the linearized equation system becomes a first-order partial differential equation system with constant coefficients, i.e.

$$\rho \frac{\partial \bar{\mathbf{v}}}{\partial t} - \operatorname{div}(\bar{\mathbf{T}}) = 0 \tag{20}$$

$$\frac{\partial \bar{\mathbf{T}}}{\partial t} - \mathbf{A}^0 \bar{\mathbf{T}} - \mathbf{B}^0 \bar{\mathbf{D}} = 0$$

where

$$\mathbf{A}^0 = \frac{\partial \mathbf{h}}{\partial \mathbf{T}}(\mathbf{T}_0, \mathbf{D}_0) \quad \text{and} \quad \mathbf{B}^0 = \frac{\partial \mathbf{h}}{\partial \mathbf{D}}(\mathbf{T}_0, \mathbf{D}_0). \tag{21}$$

The constitutive equation (2), which is represented through a nonlinear tensor-valued function $\mathbf{h}(\cdot, \cdot)$ in $\mathbf{D}$ admits only a directional linearization defined in (21). The relation (21) is obtained for the direction $\mathbf{E}_{D_0}$ defined as

$$dir(\mathbf{D}_0) = \frac{\mathbf{D}_0}{\|\mathbf{D}_0\|} = \mathbf{E}_{D_0} \tag{22}$$

of the homogeneous stretching $\mathbf{D}_0$. The Euclidean norm of the homogeneous stretching is obviously assumed to be different from zero.

To simplify the discussion we can use following substitution:

$$\mathbf{u} = \left\{ \begin{array}{c} \bar{\mathbf{v}} \\ \bar{\mathbf{T}} \end{array} \right\} = \{\bar{u}_1, ..., \bar{u}_n\}^T \tag{23}$$

and after some simple derivations one obtains the equivalent equation system written in 2D, what usually is the case, i.e.

$$\mathbf{R}\frac{\partial \bar{u}}{\partial t} + \mathbf{W}\frac{\partial \bar{u}}{\partial x_1} + \mathbf{Z}\frac{\partial \bar{u}}{\partial x_2} = \mathbf{f} \tag{24}$$

where $\mathbf{R}, \mathbf{W}, \mathbf{Z}$ are the martices with constant elements.

The above differential equation system (24) can now be checked on its type, which is the crusial information on the well-posedness requirement and correct posed boundary conditions.

In order to do that the characteristic equation

$$\det(r\mathbf{R} + w\mathbf{W} + z\mathbf{Z}) = 0 \tag{25}$$

must be solved.

Type of the differential equations system (20) depends on the roots r_i of the characteristic equation (25) obtained for arbitrary values of w and z under condition $w^2 + z^2 \neq 0$.

The equilibrium system (24) is of

- hyperbolic type for n-different and real roots r_i,

- elliptic type if there is lack of real roots r_i and

- parabolic type if r_i is n-onevalued and real root.

This condition can be checked at each Gauss-point of the defined solid volume. In this way one is able to control locally type of the governing equation system. In fact this is not complcated mathematical task which can be solved using existinf library procedures.

6 DISCUSSION OF THE SOLUTION

Since the differential equations system (20)

$$\rho \frac{\partial \bar{\mathbf{v}}}{\partial t} - \mathrm{div}(\bar{\mathbf{T}}) = 0$$

$$\frac{\partial \bar{\mathbf{T}}}{\partial t} - \mathbf{A}^0\bar{\mathbf{T}} - \mathbf{B}^0\bar{\mathbf{D}} = 0$$

is the linear equation system with constant coefficients then the solution may be obtained as a superposition of exponential solutions, [8], e.g.

$$\begin{aligned} \bar{\mathbf{v}}(t, \mathbf{x}) &= \hat{\mathbf{v}} \cdot i \cdot e^{i(\xi, \mathbf{x}) + \lambda(\xi) \cdot t} \\ \bar{\mathbf{T}}(t, \mathbf{x}) &= \hat{\mathbf{T}} \cdot e^{i(\xi, \mathbf{x}) + \lambda(\xi) \cdot t} \end{aligned} \tag{26}$$

where $\hat{\mathbf{v}}, \hat{\mathbf{T}}$ and λ are constants to be determined.

6.1 An example

Following [9] and introducing (26) in (20) one obtains an eigenvalue problem, i.e.:

$$\mathbf{S} \left\{ \begin{array}{c} \hat{\mathbf{v}} \\ \hat{\mathbf{T}} \end{array} \right\} = \left[\begin{array}{cc} \mathbf{0} & \frac{1}{\rho}\mathbf{K} \\ -\mathbf{B}^0\mathbf{M} & \mathbf{A}^0 \end{array} \right] \left\{ \begin{array}{c} \hat{\mathbf{v}} \\ \hat{\mathbf{T}} \end{array} \right\} = \lambda \left\{ \begin{array}{c} \hat{\mathbf{v}} \\ \hat{\mathbf{T}} \end{array} \right\}. \tag{27}$$

or:

$$\det(\mathbf{S} - \lambda \mathbf{I}_5) = 0. \tag{28}$$

or finally:

$$p(\lambda) = \lambda^5 + a_1\lambda^4 + a_2\lambda^3 + a_3\lambda^2 + a_4\lambda + a_5 \tag{29}$$

where: a_1, a_2, a_3, a_4, a_5 are the real-valued functions. The matrices $\mathbf{S}, \mathbf{M}, \mathbf{K}$ are specific algebraic analogs to the respectively parts of the equation system (20), for details s. [9]. This example shows how to derive some specific conditions, which allow monitoring of the numerical evaluation of the equilibrium system.

Moreover, such conditions may deliver the user important information, in this case about the linear well-posedness of the problem. One of the results obtaines within this analysis, could be formulated as a numerical statement in form of the following theorem.

Theorem 4 *The partial differential equations (20) are lineary well-posed if and only if the following inequalities are fulfilled:*

$$\mathrm{tr}(\mathbf{B}) > 0$$
$$B_{33}\sqrt{B_{11}B_{22}} > \det(\mathbf{B}) - \frac{1}{2}(B_{12} + B_{21}) \tag{30}$$
$$B_{33}\sqrt{B_{11}B_{22}} > -\det(\mathbf{B}) + \frac{1}{2}(B_{12} + B_{21}).$$

The proof of the above condition can be found in [9].
The above conditions can be easily implemented in FE analysis.

In order to check solution stability of the linearized equations system (20) we recall the characteristic polynomial (29) of the eigenvalue problem which characterizes the system. We claim that the exponential solutions of the linear equation system (20) is asymptotically stable. The necessary condition for the asymptotically stable solutions requires that all the real parts of the roots of the characteristic equation (29) must be negative. In order to check the above requirement Hurwitz criterion can be utilized and therefore the principal Hurwitz subdeterminants D_i (por. (15)) of the following matrix $\mathbf{W}$:

$$\mathbf{W} = \left[\begin{array}{ccccc} a_1 & 1 & 0 & 0 & 0 \\ a_3 & a_2 & a_1 & 1 & 0 \\ a_5 & a_4 & a_3 & a_2 & a_1 \\ 0 & 0 & a_5 & a_4 & a_3 \\ 0 & 0 & 0 & 0 & a_5 \end{array} \right] \tag{31}$$

must satisfy the inequalities

$$\forall i \in (1, 2, 3, 4, 5) \qquad D_i > 0 \tag{32}$$

7 CONCLUSIONS

The above presented analysis allows to state some conclusions.

1. A great advantage of the original hypoplastic model is its simplicity.

2. One can easy show several shortcomings caused by this simplicity but methods to circumvent the related problems are still worked out.

3. From theoretical point of view governing differential problem must be well-posed. Two conditions are most important, in particular, namely uniqueness and stability of the solution.

4. From the above analysis it can be seen that this intricate phenomenon of bifurcation can be well explained in the realm of hypoplasticity using the known theorems of bifurcation theory.

5. If one uses linearization technique for controlling the numerical evaluation of the nonlinear solution, then conditions for equivalence, i.e. homeomorphism of the linear solution and the non-linear one must be proved.

6. The dynamism of simulations can sometimes be a source of fictitious behaviors, which are not easy to separate from the characteristic ones, i.e. those which are realy written down in the nonlinear system.

7. Another point of a well-posed numerical problem is that before the analysis is started one has to know the way around which one behavior resulting from the system is expecting – which range of the parameters change and the initial values one is relevant in viewpoint of the solution in question.

8. In this context one has to underline importance of possible ,,theoretical" formulas which after carefully analysis can be easy used in the numerical applications, however those formulas mentioned in this paper are important before bifurcation point.

9. Our remarks made in this paper are addressed in the first line to those who discovered that something goes wrong in their calculation.

APPENDIX: DEFENITION OF STABILITY AT A FIXPOINT

The following nonolinear differential equation system is in question:

$$\dot{\mathbf{T}} = \mathbf{h}(\mathbf{T}) \quad \text{at} \quad M \subset R^n$$
$$\mathbf{T}(t_0) = \mathbf{T}_{t_0}$$

$$(33)$$

and one assummes that

$$\mathbf{h} : M \longrightarrow R^n \tag{34}$$

which represents a C^1-tensor(vector) field on the subspace M. The letter of power n indicates the rank of the space. Fixpoints $\mathbf{T}_0$, for which $\mathbf{h}(\mathbf{T}_0) = \mathbf{0}$, are of relevance because with them constant solutions with time $\mathbf{T}(t) \equiv \mathbf{T}_0$ are joined such solutions are to be called equilibrium states. An intiutive example of a ball on a wavy surface shows, that the solutions with initial value $\mathbf{T}_{t_0}$ in the neighbourhood of $\mathbf{T}_0$ can reveal different behaviors of stability, s. Fig. 1. The most important cases in physical viewpoint are those

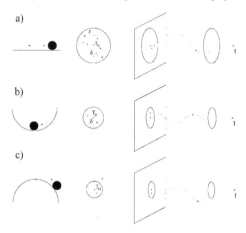

Figure 1: Stability of a fixpoint a ball in a wavy surface (left) and its abstract definition: a) stable, b) asymptotic stable, c) unstable

stable fixpoints which make negligible any small disturbances with time.

Definition 1 a) *A fixpoint $\mathbf{T}_0$ is stable (due to small perturbations in Ljapunov sense), when for all $\varepsilon > 0$ there exists $\delta > 0$ that for the solutions*

$$\mathbf{T}(t) = \mathbf{T}\left(t, \mathbf{T}(t_0)\right) = \mathbf{T}(t, \mathbf{T}_{t_0}) \tag{35}$$

the following expression holds

$$\| \mathbf{T}(t) - \mathbf{T}_0 \| < \varepsilon \quad \forall t \geq t_0, \quad \forall \mathbf{T}(t_0) = \mathbf{T}_{t_0} \quad with \quad \| \mathbf{T}(t_0) - \mathbf{T}_0 \| < \delta \tag{36}$$

(and $\mathbf{T}(t)$ for all $t \geq t_0$ exists).

b) *A fixpoint is asymptotically stable, when aditionally holds*

$$\lim_{t \to \infty} \| \mathbf{T}(t_0) - \mathbf{T}_0 \| = \mathbf{0} \quad \forall \mathbf{T}(t_0) \quad with \quad \| \mathbf{T}(t_0) - \mathbf{T}_0 \| < \delta \tag{37}$$

c) *A fixpoint which is not stable is unstable.*

The above definition does not dependent on the coordinate system used as well as the selected norm.

ACKNOWLEDGMENT

The authors are indebted to the Polish Scientific Research Committee for the financial support. A part of this project is worked out within grant No. 1578/T07/2001/20 .

REFERENCES

[1] S.S. Antman. *Bifurcation problems for nonlinear elastic structures*. Academic Press, 1977. in Applications of Bifurcation Theory, ed. Rabinowitz.

[2] S.N. Chow and J.K. Hale. *Methods of bifurcation theory*. Springer Verlag, New York, 1982.

[3] Ph. G. Ciarlet. *The finite element methods for elliptic problems*. North-Holland, Amsterdam, New York, Oxford, 1978.

[4] G. Gudehus. A comprehensive equation of state for granular materials. *Soils and Foundations, Jap. Soc. Soil Mech.*, 1993. submitted for publication.

[5] P. Hartman. *Ordinary differential equations*. John Wiley and Sons, New York, 1964.

[6] H. B. Keller and W. F. Langford. Iterations, perturbations and multiplicities for nonlinear bifurcation problems. *Arch. Rational Mech. Anal.*, 48:83–108, 1972.

[7] D. Kolymbas. A novel constitutive law for soils. In S. C. Desai, editor, *2nd Int. Conf. on Constitutive Laws for Engn. Materials*, pages 319–323, Tucson, 1987.

[8] D. G. Schaeffer. Instability and ill–posedness in the deformation of granular materials. *Int. Journ. Numer. Anal. Methods Geomech.*, 14:253–278, 1990.

[9] Z. Sikora. *Hypoplastic flow of granular materials — A numerical approach*. Technical University of Karlsruhe, Karlsruhe, 1992. Veröffentlichungen des Institutes für Bodenmechanik und Felsmechanik der Universität Fridericiana in Karlsruhe, Heft 123.

[10] B. Simpson. Engineering needs. In *Preprints of 2. International Symposion on Pre-failure Deformations Characteristics of Geomaterials*, pages 142–158, IS Torino, 1999. Jamiolkowski, M., Lancellotta, R. and Lo Presti, D.

[11] S. L. Sobolew. *Einige Anwendungen der Funktionalanalyse auf Gleichungen der mathematischen Physik*. Akademie-Verlag, Berlin, 1964.

[12] S. L. Sobolew. *Partial differential equations of mathematical physics*. Pergamon Press, Oxford, London, Edinburgh, New York, Paris, Frankfurt, 1964.

[13] W. Wu and Z. Sikora. Localized bifurcation in hypoplasticity. *Int. J. Engng. Sci.*, 29:195–201, 1990.

[14] Wei Wu. *Ein mathematisches Modell der konstitutiven Beziehungen für granulare Materialien*. PhD thesis, IBF, Technical University of Karlsruhe, 1993. Veröffentlichungen des Institutes für Bodenmechanik und Felsmechanik der Universität Fridericiana in Karlsruhe, Heft 127.

2. Instabilities and discrete systems

Bifurcations & Instabilities in Geomechanics, – Labuz & Drescher (eds.)
© *2003 Swets & Zeitlinger, Lisse, ISBN 90 5809 563 0*

Material instability with stress localization

J.D. Goddard
Department of Mechanical and Aerospace Engineering
Univertsity of California, San Diego, La Jolla, CA, USA

ABSTRACT This paper is concerned with a possible material instability arising from a non-unique dependence of stress on deformation and leading to localization of stress in otherwise homogeneous deformations. A recently published work[3] cites several constitutive theories and certain experimental evidence indicating this type of constitutive behavior, and it provides an analysis of quasi-static bifurcation of in the steady extensional deformation of certain idealized isotropic elastic solids and viscoelastic fluids. The principle objective is to raise the question as to whether the ubiquitous "force chains" observed in experiments and computer simulation of non-cohesive granular media represent stress localization. As a prelude to a more complete investigation, consideration is given to the question of the statistical distribution of contact forces in sphere assemblies. The maximum-entropy estimator of statistical thermodynamics is employed to derive the statistical distribution of contact forces in a static assembly of nearly-rigid grains. However, instead of the usual constraint of stationary energy, stationarity of stress is assumed. It is shown that under fairly general circumstances one obtains a distribution with exponential tail in contact-force magnitude, which is found in various experiments and numerical simulations. This result is exact for frictionless monodisperse sphere assemblies subject to isotropic confinement. According to the present model, the exponential tail does not depend on special models of "force propagation" in granular assemblies, of the type postulated in the contemporary physics literature. According to the above principle, the precise form of the probability density for force depends on the weight (a priori probability) assigned to elementary volumes in the state-space of contact forces. Various weight functions are discussed and the resulting distributions are compared to experiment and simulation. In its present form, the analysis given here does not apply to the "two-phase" structure associated with force chains in granular media.

1 INTRODUCTION

The concept of material constitutive instability is firmly established in continuum mechanics, particularly in the geomechanics, where it is manifest experimentally by various modes of localized deformation and failure such as necking, shear banding and localized crushing. Such instability, the analog of phase transition in thermostatics, is associated with loss of static ellipticity in the field equations and is generally due to non-convexity arising from strain softening or other non-monotone stress-deformation behavior.

This is illustrated by Fig. 1, where Figs. 1(b)-(c) allow for coexistence of different states or "phases", with possible of hysteretic transition. Fig. 1(c), which encapsulates the microbuckling instabilities associated with various foamed materials, is qualitatively different from Figs.1 (a)-(b), in that it allows for the possibility of *stress localization*, the subject of concern here which is discussed in more detail in a separate journal publication[3].

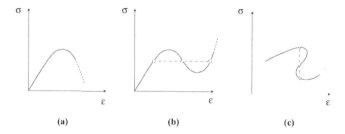

(a)　　　　　　　　(b)　　　　　　　　(c)

Figure 1: Non-monotone stress. (a) Ductile failure. (b) Strain softening with coexistent strains. (c) Catastrophic stress jumps. After [3].

There are numerous examples of various microstructural buckling instabilities in foamed and fiber-reinforced solids and in liquid crystalline and micellar systems [3], which should also be manifest in continuum-level behavior. By simply exchanging the axes in Fig. 1 one achieves in Fig. e a schematic illustration of the kind of instability envisaged ([3]. In particular, Fig. 2 (a)-(b) illustrate stress blowup in steady elongational flow, of a type found theoretically for various models of liquid-solid suspensions and polymer solutions. Also, in an article overlooked [1] by the Author [3], Prager suggested several years ago the possibility of a related phenomena in plasticity ([10]). Fig. 2(c) indicates the possibility of coexistent phases having different stress and strain components.

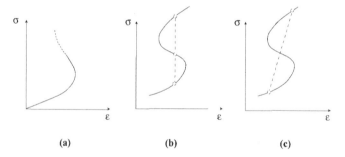

(a)　　　　　　　　(b)　　　　　　　　(c)

Figure 2: (a) Stress blowup, and stress jumps with coexistent states having (b) the same, and (c) different states of deformation. After [3]

With this brief phenomenological overview, we turn to the issue of the the granular force chains shown in Fig. 3. We recall that the two-dimensional numerical simulations of Radjai [11] on frictional disks indicates a two-phase structure, involving particles in force chains imbedded in a matrix of more or less isotopically loaded particles. This suggests a possible continuum-level description involving stress localization, possibly related to the particle-chain buckling suggested some time ago by the present author [2].

Setting aside an obvious reservation as to the applicability of continuum models to a patently micro-scale or meso-scale phenomenon, we turn to the now to a discussion of the statistics of contact forces in assemblies of non-cohesive particles.

---

[1]and pointed out to him by Prof. Zenon Mroz.

58

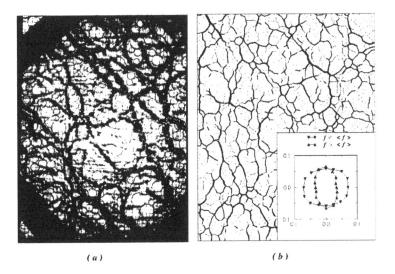

(a)   (b)

Figure 3: Contact-force chains. (a) Experimental observations on an assembly of photoelastic disks. After [1]. Courtesy of A. Drescher. (b) Numerical simulation, showing magnitudes of above- (heavy lines) and below-average (light lines) forces, and their mean orientation or "fabric" (inset). After [11]. Courtesy of F. Radjai.

2 A STATISTICAL-THERMODYNAMIC MODEL OF CONTACT FORCES

We begin by a brief recapitulation of the statistical-thermodynamics essentials [6] for an arbitrary mechanical system involving a presumably large number of mechanical degrees of freedom. Thus, we let $\mathbf{z} \in \Omega$ denote a representative point in the relevant state ("phase") space Ω, endowed with probability measure $P(\mathbf{z})d\Omega(\mathbf{z})$, where $d\Omega(\mathbf{z})$ is an elemental state-space measure that remains to be defined. Statistical averages (expectations) of various mechanical variables $\Lambda(\mathbf{z})$ are given by

$$\langle \Lambda \rangle = \int_\Omega \Lambda(\mathbf{z})P(\mathbf{z})d\Omega(\mathbf{z}), \tag{1}$$

including the special case $\langle c \rangle = c$ for constant c, with normalization of P implied by c=1
.

The standard statistical-thermodynamical estimate for the unknown probability distribution $P(\mathbf{z})$ requires stationarity of the (entropy) functional

$$\langle \log P \rangle = \int_\Omega P(\mathbf{z}) \log P(\mathbf{z})d\Omega(\mathbf{z}) \tag{2}$$

under arbitrary variation of $P(\mathbf{z})$ and subject to a discrete set of constraints on mechanical variables of the form

$$< \Lambda_i >= \text{const.}, \quad i = 1, 2, \ldots \tag{3}$$

This results in the *canonical* distribution

$$P(\mathbf{z}) = Z^{-1} \exp\{-\sum_i \beta_i \Lambda_i(\mathbf{z})\} \tag{4}$$

where

$$Z = \int_{\Omega} \exp\{-\sum_i \beta_i \Lambda_i(\mathbf{z})\} d\Omega(\mathbf{z}), \quad \text{with } \langle \Lambda_i \rangle = -\frac{\partial}{\partial \beta_i} \log Z, \tag{5}$$

is the *partition function*, and the β_i represent Lagrange multipliers corresponding to distinct "temperatures" β_i^{-1}.

In the classical thermodynamic setting, [6], $\mathbf{z}$ represents a set of generalized coordinates and momenta $\{q_k, p_k\}$, with dynamically invariant measure $d\Omega(\mathbf{z}) = \prod_k dq_k dp_k$ in the associated phase space. The celebrated Boltzmann distribution arises from (3) as the special case of a single invariant $\Lambda = U$, the internal energy.

There are obvious reservations concerning the application of the above formalism to static granular assemblies. First of all, the randomness in granular systems is due principally to spatial disorder associated with a random static packing rather than the chaotic thermal motion that allows thermodynamic systems to explore all regions of phase space.

Setting aside the preceding reservation, it is evident that a straight-forward application of above the entropy principal would yield something akin to the Boltzmann distribution for assemblies of elastic particles subject to the constraint of stationary elastic strain energy [9]. For example, in the case of nearly rigid, non-cohesive frictionless elastic spheres with elastic contact energy f^ν given in terms of normal compressive contact force $f \geq 0$, the above principle yields a canonical distribution of the form

$$P(f) = Z^{-1} \exp(-\beta f^\nu) \tag{6}$$

However, the probability density in f, say, $\Phi(f)$, such that

$$\Phi(f)df = P(f)d\Omega(f), \tag{7}$$

obviously depends on the on state-space measure $d\Omega(f)$, that is on the *a priori* weight assigned to the interval df in the state-space of contact forces. Since elastic energy is conserved in any admissible quasi-static rearrangement of particles, it seems appropriate to identify the state-state measure with this energy, such that

$$d\Omega = cf^{\nu-1}df, \quad c = \text{const.} \tag{8}$$

For later reference, we note the values

$$\nu = \begin{cases} 1, & \text{linear elastic contact} \\ 3/2, & \text{empirical elasticity} \\ 5/3, & \text{Hertzian contact}[2, 9] \end{cases} \tag{9}$$

where the empirical value is that inferred from the pressure dependence of elastic stiffness or wave speed [2], which may actually involve a variation number of contacts with imposed strain. Note that the linear elastic contact gives rise to a Gaussian distribution for $\Phi(f)$ in (7).

In the limit of perfectly rigid particles the strain-energy principle is no longer appropriate, and we explore an alternative based on minimization of (2) subject to a given granular (pressure) stress

$$\mathbf{P} = n_c \langle \mathbf{f} \otimes \mathbf{r} \rangle, \tag{10}$$

where n_c is contact density (number per unit volume), $\mathbf{f}$ is vectorial contact force and $\mathbf{r}$ is the so-called branch vector that connects centroids of adjacent grains. In (10) we have written the standard expression for static stress in a homogeneous granular assembly [2, 5] in terms of a statistical average over intergranular contacts, to be obtained from the probability distribution for $\mathbf{z} \equiv \{\mathbf{f}, \mathbf{r}\}$.

The minimization (2) of subject to stationarity of (10) yields the appropriate version of (4) as a joint probability distribution

$$P(\mathbf{f}, \mathbf{r}) = Z^{-1} \exp\{-\mathbf{B} : \mathbf{P}\}, \quad \text{where } \mathbf{B} : \mathbf{P} = n_c \mathbf{f} \cdot \mathbf{B} \cdot \mathbf{r}, \tag{11}$$

and

$$Z(\mathbf{B}) = \int_{\mathbf{f}} \int_{\mathbf{r}} \exp\{-n_c \mathbf{f} \cdot \mathbf{B} \cdot \mathbf{r}\} d\Omega(\mathbf{f}) d\Omega(\mathbf{r}), \quad \text{with } \mathbf{P} = -\frac{\partial}{\partial \mathbf{B}} \log Z. \tag{12}$$

The last two relations give Z as a well-defined function $Z(\mathbf{B})$ of (Lagrange multiplier) $\mathbf{B}$ and, hence, under a mild assumption of invertibility, give $\mathbf{B}$ as a function of the applied stress $\mathbf{P}$

The above provides a general theory for assemblies of grains of arbitrary shape and size subject to arbitrary states of stress. Certain simplifications arise for the important special case of small deformations from statistically well-defined assemblies of grains, since we may assume that the probability distribution of $\mathbf{r}$ is that associated with the initial state, say $P_0(\mathbf{r})$. In that case, the above equations can be replaced by a simpler set involving the conditional probability $P(\mathbf{f}|\mathbf{r})$. However, in order to obtain more definite results, we consider the further special case of initially random isotropic assemblies of equi-sized frictionless spheres of diameter d.

2.1 *Isotropic monodisperse assemblies of frictionless spheres*

$$\mathbf{f} = f\mathbf{e}, \quad \text{where } \mathbf{e} = \mathbf{r}/r, \ r = |\mathbf{r}| \tag{13}$$

and

$$P_0(\mathbf{r}) = \frac{1}{4\pi d} \delta(r - d), \quad \text{with } d\Omega(\mathbf{r}) = r dr d\Omega(\mathbf{e}), \ d\Omega(\mathbf{e}) \equiv \sin\theta d\theta d\phi \tag{14}$$

with f representing compressive contact force, $\mathbf{e}$ the unit contact normal, θ, ϕ polar coordinates on the unit sphere and δ the Dirac delta. Hence, one finds from the general result (11) that

$$P(f) = Z^{-1} \int_{\Omega_s} \exp\{-\beta(\mathbf{e})f\} d\Omega(\mathbf{e}), \tag{15}$$

with

$$Z(\mathbf{B}) = \int_0^\infty \int_{\Omega_s} \exp\{-\beta(\mathbf{e})f\} d\Omega(\mathbf{e}) d\Omega(f), \tag{16}$$

where Ω_s denotes the surface of the unit sphere, and

$$\beta(\mathbf{e}) = \beta(\mathbf{e}, \mathbf{B}) = n_c d \, \mathbf{e} \cdot \mathbf{B} \cdot \mathbf{e} \tag{17}$$

Now, given a specific form for $d\Omega(f)$, it is clear that Z represents an isotropic scalar-valued function of the second-rank tensor $\mathbf{B}$ and its derivative $\partial Z/\partial \mathbf{B}$ is a tensor-valued isotropic function of $\mathbf{B}$. Hence, the derivative in (12) defines an isotropic tensor function $\mathbf{B}(\mathbf{P})$.

For example, given the power-law form (8), one finds that

$$Z = c\Gamma(\nu) \int_{\Omega_s} \beta^{-\nu}(\mathbf{e}) d\Omega(\mathbf{e}) \tag{18}$$

where Γ denotes the gamma function.

It appears possible to derive explicit analytic forms for $Z(\mathbf{B})$ in terms of the eigenvalues or principal isotropic invariants of $\mathbf{B}$ for certain special values of ν and/or special symmetries of $\mathbf{B}$, e.g. such as transverse isotropy (orthotropy), symmetries which are identical with those of the stress $\mathbf{P}$. In particular, for symmetric stress $\mathbf{P} = \mathbf{P}^T$, the tensor $\mathbf{B}$ is also symmetric and (17) can be reduced to the simple form

$$\beta(\mathbf{e}) = n_c d\{(\lambda_1 \cos^2 \phi + \lambda_2 \sin^2 \phi) \sin^2 \theta + \lambda_3 \cos^2 \theta\} \tag{19}$$

where the polar angles θ, ϕ are referred to the principal axes of $\mathbf{B}$, whose associated characteristic values are denoted here by λ_i

For the present purposes, we further specialize to the case of confinement by an isotropic pressure, where $\mathbf{P}$ and, hence, $\mathbf{B}$ are proportional to the unit tensor.

2.2 Isotropic confinement

In the case of isotropic $\mathbf{B}$, the quantity β in (15) is independent of $\mathbf{e}$. Hence, with a bit of algebra, one can eliminate various constants and express (15) in the form

$$P(f) = Z^{-1} \exp\{-\beta f\}, \tag{20}$$

with

$$Z = \int_0^\infty \exp\{-\beta f\} d\Omega(f), \tag{21}$$

from which β is given implicitly in terms of the confining pressure or, equivalently, in terms of the mean of f by

$$\langle f \rangle = \beta^{-1} \frac{\int_0^\infty e^{-s} s \, d\Omega(s/\beta)}{\int_0^\infty e^{-s} d\Omega(s/\beta)}, \tag{22}$$

once the functional form of $d\Omega(f)$ is specified. For example, with the power law (8), one readily finds that $\beta = \nu \langle f \rangle^{-1}$ and, hence, that the density (7) is given by the Poisson distribution

$$\Phi(F) = \nu \frac{(\nu F)^{\nu-1}}{\Gamma(\nu)} e^{-\nu F}, \quad \text{where } F = \frac{f}{\langle f \rangle} \tag{23}$$

Although our derivation supposes no detailed expression for redistribution of forces, this result is identical with the mean-field solution for a specific model of "force propagation" [7]. We consider now a further comparison with other work.

3 COMPARISON WITH EXPERIMENT AND SIMULATION

Mueth et al. [8] employ the empirical expression

$$\Phi(F) = \alpha(1 - \beta e^{-F^2}) \exp\{-\nu F\} \tag{24}$$

to fit their experimental data on sphere assemblies. Fig. 4 compares (a) the curve obtained from their values

$$\beta = 0.75, \quad \nu = 1.5$$

and with $\alpha(\beta, \nu)$ given by normalization of (24), (b) the numerical simulations of Radjai et al. [11] on disks, employing the values

$$\beta = 0.6, \quad \nu = 1.35,$$

they propose for the same the empirical formula (24), and (c) the distribution (23) with $\nu = 3/2$. The latter clearly fails to capture the behavior near $f = 0$, as does any model of the type (23) with $\nu > 1$.

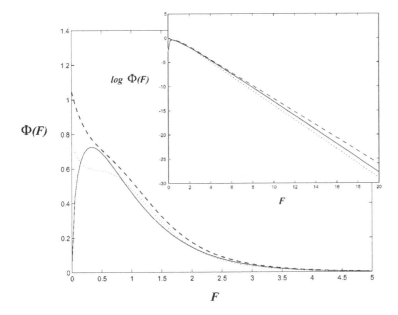

Figure 4: Comparison of contact force distributions: (a) Dotted curve ..... , empirical fit of experiments on spheres, (b) Dashed curve _ _ _ _, numerical simulations on disks [11],(c) Solid curve ____, Eq. (23) with $\nu = 3/2$. The inset semi-log plot shows exponential tails.

Not shown in Fig. 4 are the corresponding results the "q model" of Coppersmith et al. [7, 13, 9, 12], a stochastic load transfer model of a type which goes back at least to the early diffusional model of Harr [4]. The two-dimensional version of this model gives a distribution of the form (23) with $\nu = 2$ [9], with a correspondingly large departure from experiment near $f = 0$ and a more rapidly decay of the exponential tail. But for the presumably accurate numerical simulations of Radjai et al. [11], departures near $f = 0$ might be attributed to experimental error [7, 9]. While the present treatment can in principle eliminate such discrepancies by adjusting the the assumed measure $d\Omega$ near $f = 0$, there is no evident theoretical justification for doing so.

4 CONCLUSIONS

The above considerations suggest that static force distributions in random assemblies of nearly rigid particles may be as well described by statistical-thermodynamics methods as by specific models of stochastic force distribution. Among the open questions concerning the present model is the assignment of *apriori* probability to different regions of state space. The tentative interpretation in terms of elastic contact energy might be based on the argument that systems of nearly rigid particles adjust forces globally according to the applied stress, somewhat independently of a simultaneous local minimization of elastic contact energy. At any rate, further work is needed to extract results from the present model for the case of non-isotropic confinement, actually employed in the numerical simulations of Radjai et al. [11]. Should further, more detailed experiment and numerical simulation tend to support the present approach, an extension to two-phase structure associated with force chains would be most interesting.

5 ACKNOWLEDGEMENT

Partial support from National Aeronautics and Space Administration Grant NAG3-2465 is gratefully acknowledged.

REFERENCES

[1] A. Drescher and G. De-Josselin-de Jong. Photoelastic verification of a mechanical model for the flow of a granular material. *J. Mech. Phys. Solids (UK)*, 20(5):337–51, 1972.

[2] J. D. Goddard. Nonlinear elasticity and pressure-dependent wave speeds in granular media. *Proc. R. Soc. Lond. A, Math. Phys. Sci. (UK)*, 430(1878):105–31, 1990.

[3] J. D. Goddard. Material instability with stress localization. *Journal of Non-Newtonian Fluid Mechanics*, 102:251–61, 2002.

[4] M. E. Harr. *Mechanics of particulate media: a probabilistic approach*. McGraw-Hill, New York, 1977.

[5] Hans J. Herrmann, J. P. Hovi, and S. Luding. *Physics of dry granular media*. Kluwer Academic, Dordrecht ; Boston, Mass, 1998.

[6] T. L. Hill. *An introduction to statistical thermodynamics*. Addison-Wesley, Reading, Mass., 1960.

[7] C. H. Liu, S. R. Nagel, D. A. Schecter, S. N. Coppersmith, S. Majumdar, O. Narayan, and T. A. Witten. Force fluctuations in bead packs. *Science*, 269(5223):513–15, 1995.

[8] D. M. Mueth, H. M. Jaeger, and S. R. Nagel. Force distribution in a granular medium. *Phys. Rev. E*, 57(3):3164–9, 1998.

[9] M. L. Nguyen and S. N. Coppersmith. Properties of layer-by-layer vector stochastic models of force fluctuations in granular materials. *Phys. Rev. E*, 59(5):5870–80, 1999.

[10] W. Prager. On ideal locking materials. *Transactions of the Society of Rheology*, 1:169–75, 1957.

[11] F. Radjai, D. E. Wolf, M. Jean, and J. J. Moreau. Bimodal character of stress transmission in granular packings. *Phys. Rev. Lett.*, 80(1):61–4, 1998.

[12] J. H. Snoeijer and J. M. J. van Leeuwen. Force correlations in the q model for general q distributions. *Phys. Rev. E*, 65(5):051306/1–12, 2002.

[13] J. E. S. Socolar. Average stresses and force fluctuations in noncohesive granular materials. *Phys. Rev. E*, 57(3):3204–15, 1998.

Bifurcations & Instabilities in Geomechanics, – Labuz & Drescher (eds.)
© 2003 Swets & Zeitlinger, Lisse, ISBN 90 5809 563 0

Continuous and discrete modeling of failure modes in geomechanics

F. Darve, G. Servant
Laboratoire Sols Solides Structures, RNVO, INPG, France

F. Laouafa
INERIS, Verneuil en Halatte, France

ABSTRACT: After recalling briefly the main results obtained in the framework of the continuum mechanics by applying Hill's condition of stability to loose and dense sands, an analysis of grain avalanches by a discrete form of the second order work is proposed. It is shown some spatial and temporal correlations between the bursts of kinetic energy and the negative values of the discrete second order work. While localized modes of failure with shear band formation are well described by the vanishing values of the determinant of the acoustic tensor, it is conjectured in conclusion that diffuse modes of failure might be characterized by the vanishing values of the second order work (what corresponds for conventional elasto-plastic models to the vanishing values of the determinant of the symmetric part of the constitutive matrix).

1 INTRODUCTION

In the framework of non-associated elasto-plasticity, it has been shown (Bigoni & Hueckel 1991, Nova 1994) that some bifurcations, instabilities and losses of uniqueness can occur strictly inside the conventional plastic limit condition (defined by the vanishing values of the determinant of the constitutive matrix). This is basically due to the non-symmetry of the constitutive matrix for non-associated materials. The stress states, where these bifurcations, instabilities and losses of uniqueness are developing, are usually known from an experimental point of view as "failure states". These failure states are linked to various modes of failure. It has been shown experimentally (Vardoulakis & Sulem 1995, Desrues 1984) that localized modes of failure appear before the plastic limit condition is reached, typically for dense granular materials. From a theoretical point of view, this kind of results can be simulated in the framework of non-associated elasto-plasticity.

For undrained loose sands, before an eventual liquefaction along an undrained triaxial loading, the deviatoric stress goes through a maximum. If we add at this maximum a "small" axial additional force, the sample exhibits a sudden brutal failure with a chaotic displacement field. This failure occurs strictly before the conventional plastic limit condition without any localization pattern. From a theoretical point of view, for this loose granular material and at these stress states, the localization criterion will not be fulfilled. This is the essential reason why this mode of failure has been called "diffuse" failure (Darve & Laoufa 2000). If Hill's condition of stability is considered, it appears that the second order work is vanishing at the deviatoric stress peak (Darve & Chau 1987). Thus Hill's condition of stability has been applied in a systematic manner to loose and dense sands, and the main results have been recalled in section 2 of this paper.

Granular avalanches have been studied very extensively by physicists of granular media (see for example, Hermann, Havi & Luding 1998). These are indeed peculiar events. The slope at rest can be considered as a solid, while, if the angle of the slope is increasing, there is a brutal phase change and suddenly the sand is flowing like a liquid ! These transitions have been viewed as a critical state and these phenomena have been utilized as a paradigm for the theory of "self-organized criticality" (Bak 1996).

In this context it was interesting to investigate if it is possible to analyze granular avalanches by a discrete form of the second order work. The second part of this paper is thus devoted to simulations of induced granular avalanches.

For that, a 2D granular material of Schneebeli type (constituted by wood piled cylinders) was considered. The experiments were conducted in the bi-dimensional shear apparatus "1γ2ε" (Joer et al. 1998). The discrete element method used for the computations is called "Contact Dynamics" (Moreau 1994). Basically it considers rigid grains interacting each other by coulombian friction. The granular medium and the discrete element method are presented in section 3.

Granular avalanches are characterized by simultaneous sudden motions of several grains, what means a burst of the kinetic energy. Thus the kinetic energy of each particle was computed and compared to the value of its discrete second order work.

Four different boundary conditions have been considered and the results of 2 cases will be compared and discussed in sections 4 and 5.

The close link, which is clearly exhibited by the computations, between granular avalanches and negative values of the second order work is emphasized in conclusion.

2 FAILURE MODES IN CONTINUUM GEOMECHANICS

2.1 *Drained triaxial compressions*

As the most classical case, let us consider a drained triaxial compression on a dense sand. The lateral pressure is kept constant, while the axial stress goes through a maximum.

Let us apply several criteria to the stress maximum :

- Lyapunov's definition of stability (Lyapunov 1907) :

If a small negative $d\sigma_l$ is applied, a small elastic strain response is obtained. On the contrary, the application of a small positive $d\sigma_l$ induces a large plastic strain response. Thus, according to Lyapunov's definition of stability, σ_l peak has to be considered as unstable.

- Hill's condition of stability (Hill 1958) :

The second order work is equal, in axisymmetric conditions, to :

$$d^2W = d\sigma_1 d\varepsilon_1 + 2d\sigma_3 d\varepsilon_3$$

$$= d\sigma_1 d\varepsilon_1 \qquad \text{for drained triaxial loading.}$$

As it vanishes at σ_l peak, after Hill's condition of stability this stress state is unstable.

- Plasticity criterion and flow rule :

For axisymmetric conditions a rate-independent incrementally piece-wise linear constitutive relation takes the following form :

$$\begin{pmatrix} d\sigma_1 \\ \sqrt{2}d\sigma_3 \end{pmatrix} = \mathbf{M} \begin{pmatrix} d\varepsilon_1 \\ \sqrt{2}d\varepsilon_3 \end{pmatrix} \tag{1}$$

For a constant lateral stress and at σ_l peak, it comes :

$$\begin{pmatrix} 0 \\ 0 \end{pmatrix} = \mathbf{M} \begin{pmatrix} d\varepsilon_1 \\ \sqrt{2}d\varepsilon_3 \end{pmatrix}$$

Thus, we have :
  a plasticity criterion given by : $\det \mathbf{M} = 0$ \hfill (2)
  and the flow rule corresponding to :

$$\mathbf{M} \begin{pmatrix} d\varepsilon_1 \\ \sqrt{2}d\varepsilon_3 \end{pmatrix} = \begin{pmatrix} 0 \\ 0 \end{pmatrix} \tag{3}$$

If we consider our "octolinear" constitutive relation (Darve & Labanieh 1982), both relations (2) and (3) take respectively the following expressions :

$$\begin{cases} E_1^+ E_3^+ = 0 \\ V_1^{3+} d\varepsilon_1 + d\varepsilon_3 = 0, \end{cases} \tag{4}$$

what corresponds to basic definitions of E_i and V_i^j as generalized tangent Young's moduli and Poisson's ratio.

2.2 Undrained triaxial compressions

Let us come back to the case of undrained loose sands, already considered in the introduction. Perfectly undrained loading means that the sample is deformed at constant volume. Thus :

$$d\varepsilon_v = d\varepsilon_1 + 2d\varepsilon_3 = 0 \tag{5}$$

As it is well known from the experiments, before a possible liquefaction, $q = \sigma_1 - \sigma_3$ goes through a maximum.
It is interesting to try to apply the same criteria, as for the drained triaxial compressions, to this maximum. Thus it comes :

- Lyapunov's definition of stability :
Because : $F / S = q$, where F and S are respectively the axial force and the cross section of the sample, with a "small" additional axial force, a brutal failure of the sample is developing.
q peak is thus unstable according to Lyapunov's definition.

- Hill's condition of stability :
With constraint (5), the axisymmetric second order work can be written successively :

$$d^2W = d\sigma_1 d\varepsilon_1 + 2d\sigma_3 d\varepsilon_3$$

$$= dq d\varepsilon_1$$

q peak is also unstable after Hill's condition of stability.

- Bifurcation criterion and flow rule :
With the same assumptions as previously, the constitutive relation takes the following form :

$$\begin{pmatrix} dq \\ d\varepsilon_v \end{pmatrix} = \mathbf{N} \begin{pmatrix} d\varepsilon_1 \\ d\sigma_3 \end{pmatrix} \tag{6}$$

With constraint (5) and at q peak, it comes :

$$\begin{pmatrix} 0 \\ 0 \end{pmatrix} = \mathbf{N} \begin{pmatrix} d\varepsilon_1 \\ d\sigma_3 \end{pmatrix}$$

what implies :
   a bifurcation criterion given by : $\det \mathbf{N} = 0$ \hfill (7)
   and a failure rule whose expression is :

$$\mathbf{N} \begin{pmatrix} d\varepsilon_1 \\ d\sigma_3 \end{pmatrix} = \begin{pmatrix} 0 \\ 0 \end{pmatrix} \tag{8}$$

It is interesting to remark that Equation 7 is fulfilled strictly inside the limit plastic surface (defined by Equation 2). That is the basic reason why it must be called "bifurcation criterion". Besides Equation 8 is a mixed equation relating strains and stresses and it must not be confounded with the flow rule (Equation 3).

With the notations of our octolinear constitutive relation, Equations 7 and 8 are taking the following detailed expressions :

$$
\begin{cases}
2\dfrac{E_1^-}{E_3^-}(1-V_3^{1-}-V_3^{3-})+1-2V_1^{3-}=0 \\[2mm]
E_1^- d\varepsilon_1 + \left(2\dfrac{E_1^-}{E_3^-}V_3^{1-}-1\right)d\sigma_3 = 0
\end{cases}
\tag{9}
$$

2.3 Generalisation : cones of unstable directions

The previous reasoning can be generalized by considering axisymmetric strain proportional paths defined incrementally by :

$$
\begin{cases}
d\varepsilon_1 = \text{positive constant} \\
d\varepsilon_1 = d\varepsilon_2 \ (\text{axisymmetric conditions}) \\
d\varepsilon_1 + 2Rd\varepsilon_3 = 0 \ (\text{with } R = \text{constant for a given path})
\end{cases}
\tag{10}
$$

R = 1 corresponds to the previous undrained path, while for 0 < R < 1 the paths are dilatant as a constraint.

The computations (Darve & Laouafa 2001), show that, for a certain range of values for R (which depends on the material, its initial density and the mean stress level), $(\sigma_1 - \sigma_3 / R)$ goes through a maximum.

At this maximum, Lyapunov's definition of stability can be applied and the second order work with the constraint :

$$
d\varepsilon_1 + 2Rd\varepsilon_3 = 0
$$

takes the following expression :

$$
d^2W = (d\sigma_1 - d\sigma_3 / R)d\varepsilon_1
\tag{11}
$$

which is vanishing at $(\sigma_1 - \sigma_3/R)$ peak.

By writing the constitutive relation under the following form :

$$
\begin{pmatrix} d\sigma_1 - d\sigma_3 / R \\ d\varepsilon_1 + 2Rd\varepsilon_3 \end{pmatrix} = \mathbf{Q}\begin{pmatrix} d\varepsilon_1 \\ d\sigma_3 / R \end{pmatrix}
\tag{12}
$$

the bifurcation criterion is given by :
det $\mathbf{Q} = 0$, or equivalently :

$$
2\dfrac{E_1^-}{E_3^-}(1-V_3^{3-})R^2 - 2\left(V_1^{3-}+\dfrac{E_1^-}{E_3^-}V_3^{1-}\right)R+1 = 0
\tag{13}
$$

This equation has 2 real solutions in ▶ if the following discriminant is strictly positive :

$$
-\dfrac{2}{(E_3^-)^2}\left(1-V_3^{3-}-2V_1^{3-}V_3^{1-}\right)^2 \det \mathbf{M}^s > 0
\tag{14}
$$

where $\mathbf{M}^s$ is the symmetric part of the constitutive matrix $\mathbf{M}$ defined by Equation 1.

Thus a cone of unstable stress directions is obtained when :

$$
\det \mathbf{M}^s < 0,
\tag{15}
$$

what represents the condition of "non-controllability" as proposed by Nova (1994).

68

For incrementally linear constitutive relations and for classical elasto-plastic models, the equation of the boundaries of the unstable domain is thus given by the vanishing values of the determinant of the symmetric part of the constitutive matrix (Equation 15).

The slope of the first unstable stress direction is equal to :

$$\frac{\Delta\sigma_1}{\sqrt{2}\Delta\sigma_3} = \frac{2E_1^-\left(1-V_3^{3-}\right)\left(E_1^-V_3^{1-}-E_3^-V_1^{3-}\right)}{\sqrt{2}E_3^-\left(E_1^-\left(1-V_3^{3-}-V_1^{3-}V_3^{1-}\right)-E_3^-\left(V_1^{3-}\right)^2\right)}$$

(16)

Finally the rupture rule is given by :

$$E_1^-d\varepsilon_1 + \left(2\frac{E_1^-}{E_3^-}V_3^{1-}-\frac{1}{R}\right)d\sigma_3 = 0$$

(17)

In Figure 1, the cones of unstable stress directions have been plotted for a loose and a dense sand and for the octo-linear constitutive relation just used before.

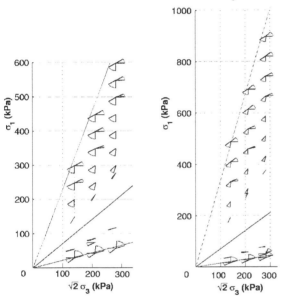

Figure 1. Cones of unstable stress directions in Rendulic plane for axisymmetric conditions in the cases of a loose sand (on the left) and of a dense sand (on the right) for octolinear constitutive model.

3 GRANULAR AVALANCHES IN DISCRETE GEOMECHANICS

Granular materials present a double nature. From many experiments (oedometric tests, triaxial tests,...) and in many practical applications (engineering works,...), they behave like continuous media. Constitutive equations and finite element codes are the usual tools to solve this class of problems.

But from other experiments (flows, segregation, erosion,...) or applications (rockfalls,...), their discrete nature can not be circumvented.

In this section and the following ones, granular materials will be considered as discrete media.

3.1 *The utilized discrete material*

This is a bi-dimensional medium (the so-called Schneebeli material) constituted by piled cylinders in wood (see Fig. 2 for a configuration with 335 grains). The friction coefficient between grains has been estimated to 0.5317. The diameters are equal to 13, 18 and 28 mm. These large dimensions allow an automatic image acquisition and data processing. From these photographic techniques, it is possible to obtain incremental displacements and rotations fields (or velocity fields) for the whole assembly of particles. The experiments have been performed inside a plane shear box (Joer et al. 1998). Four different boundary conditions have been considered. They are summarized in Figure 3. In this paper two cases will be presented : the first one (3-a), where the slope is put at failure in an active state by the displacement to the outside (the left side of Fig. 3-a) of the downstream plate, the second one (3-d), where a failure state is reached by a box rotation.

Figure 2. Typical configuration of an assembly of 335 wood piled cylinders.

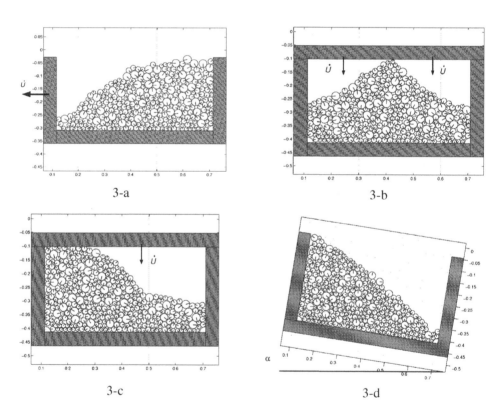

Figure 3. The four studied cases with their boundary conditions. Cases 3-a and 3-d are presented in the present paper.

3.2 *The utilized discrete element method*

This method is based on the Contact Dynamics theory developed by J.J. Moreau (1994). The numerical code, which has been utilized for the present computations, has been achieved by J. Lanier & M. Jean (1999).

The grains are assumed to be perfectly rigid and thus their are fulfilling Signorini condition as given in Figure 4 (a contact force exists only for a perfect intergranular contact $\delta = 0$). In addition to the non-penetrability condition, the interaction between grains is given by a coulombian friction. The tangential component R_t of the force is included into the range of values : $-\mu R_n$ and $+\mu R_n$, where R_n is the normal component and μ the intergranular friction coefficient (see Fig. 5).

The problem is numerically solved by taking into account the equations of dynamics for each grain. The numerical algorithm for the time integration is implicit (Lanier & Jean 1999).

This code is very different from the ones which has been previously developed by P. Cundall (1979), essentially because there are neither springs, slides nor dashpots between grains.

From the numerical experimentation it appears that the initial configuration of the grain assembly plays a very important role with respect to the model capacity to simulate realistic failure mechanisms. This is the reason why the numerical initial configurations reproduce as accurately as possible the real experimental configurations (characterized by photographic techniques). Figure 6 shows a comparison between typical experimental and numerical values of the displacement fields at the same displacements of the downstream plate.

Two conclusions can be exhibited from this comparison :

- the agreement is quite good, what illustrates the capacity of the numerical model, once more, if the initial configuration is well reproduced,
- in any case it has not been found any localized failure : in Figures 6-c and 6-e there are discontinuity lines which are separating a domain in motion from an area in repose. But, the domain in motion is neither in translation nor in rotation along some shear bands. On the contrary, from the experimental measurements as well as from the numerical simulations, the displacement field of the grains in motion has a chaotic character. Thus, in these experiments, the failure has rather a diffuse nature.

Figure 4. Illustration of the non-penetrability condition (i.e. Signorini condition).

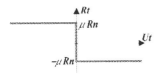

Figure 5. Illustration of the intergranular coulombian friction (μ is the intergranular friction coefficient).

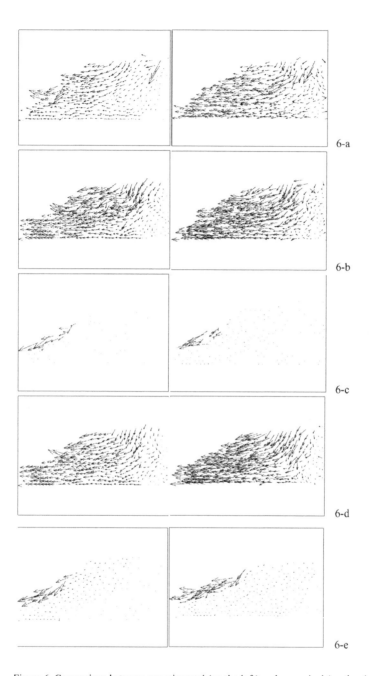

6-a

6-b

6-c

6-d

6-e

Figure 6. Comparison between experimental (on the left) and numerical (on the right) displacement fields.

72

4 SLOPE FAILURE IN AN ACTIVE STATE

In the framework of discrete mechanics the second order work has to be defined by discrete variables, as force, displacement,...

Thus we propose to consider (in 2D) :

$$d^2W = \overrightarrow{dF}.\overrightarrow{dl} + dC.d\Omega \tag{18}$$

where $\overrightarrow{dF}$ is the increment of the total force applied to a given grain, $\overrightarrow{dl}$ the increment of displacement of this grain, dC the increment of the total moment and $d\Omega$ the increment of rotation (around the cylinder axis).

The "global" second order work will be also computed by adding the values of the second order work of each particle :

$$D^2W = \sum_{k=1}^{N} d^2W_k \tag{19}$$

where d^2W_k is the second order work of particle "k" and N the number of particles.

All the increments of the various variables are defined according to an increment of the loading parameter (which is the displacement of the downstream plate).

The objective was to exhibit a possible correlation between second order work and failure. One indication in practice is the occurrence of sudden motions. Indeed granular avalanches are characterized by such sudden slides of several grains. Kinetic energy represents clearly a good way to exhibit motions.

Thus the kinetic energy of each grain has been computed by considering the usual definition :

$$e_c = \frac{1}{2}mv^2 + \frac{1}{2}I\omega^2, \tag{20}$$

where m is the mass of the particle, v its speed, I the inertia moment with respect to the cylinder axis and ω the rotation speed around this axis.

The "global" kinetic energy has been also computed :

$$E_c = \sum_{k=1}^{N} e_c^k, \tag{21}$$

where e_c^k is the kinetic energy of particle "k".

Figure 7 presents a comparison between the values of the negative second order work and the kinetic energy of each particle. A spatial correlation appears in the figure between the positions of the particles with negative values of their discrete second order work and with high values of their kinetic energy. As the emergence of kinetic energy implies local granular avalanches of more or less wide, these results show that negative values of discrete second order work could be an indicator of local failures, which are not localized in the cases considered here.

Figure 8 concerns "temporal" correlations, while Figure 7 was devoted to spatial correlations. If a loading parameter is considered (as the physical time or the displacement of the downstream plate at a constant rate), it becomes possible to follow the evolution of global quantities as the second order work (Equation 19) and the kinetic energy (Equation 21) of all particles.

Figure 8 presents these evolutions. It shows that the kinetic energy of the whole granular assembly is varying with sudden bursts which correspond to avalanches. It is possible to distinguish 5 such main bursts. As regards the evolution of the global second order work, Figure 8 exhibit very erratic variations of this quantity from positive to negative values. Each point corresponds to one computation (1080 successive steps). In these clouds of points a trend however is emerging. The peaks of negative values of the global second order work are in close correlation with the bursts of kinetic energy. Indeed it is easy to distinguish 5 peaks of negative D^2W for the same values of the loading parameter as for the 5 bursts of kinetic energy E_c.

In the next section another case, with very different boundary conditions, is presented and discussed by considering the same basic diagrams.

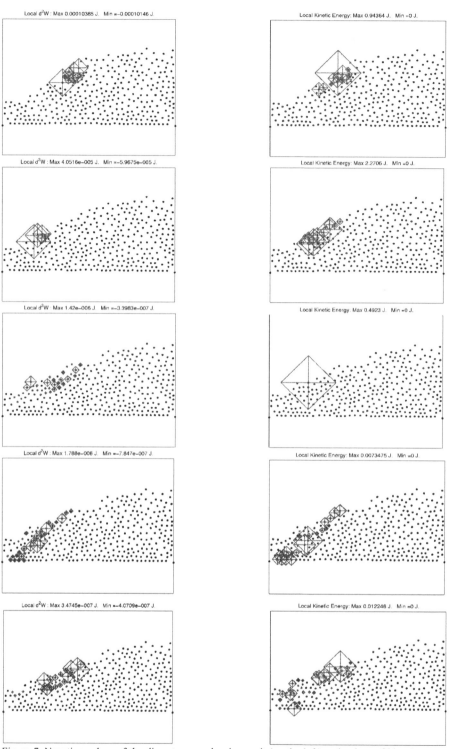

Figure 7. Negative values of the discrete second order work (on the left) and values of kinetic energy (on the right) for different plate displacements at the left of the slope.

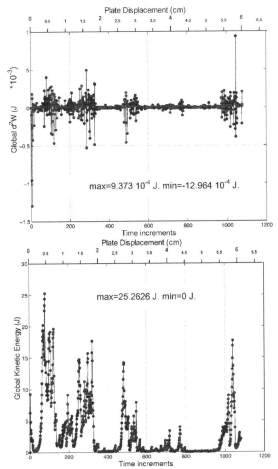

Figure 8. Variations of the global discrete second order work (at the top) and the global kinetic energy (at the bottom) versus plate displacement.

5 SLOPE FAILURE BY ROTATION

The case 3-d of Figure 3 is considered here. The slope failure is reached by continuous rotation of the box. The loading parameter is thus constituted by the angle of rotation from an initial zero value.

Figure 9 compares the spatial positions of the particles with negative discrete second order works and with positive kinetic energy. As on Figure 7, only the negative values of d^2W have been plotted. A spatial correlation appears from Figure 9. Here also the successive granular avalanches are characterized not only by bursts of kinetic energy but also by negative values of discrete second order work. The very different values of kinetic energy of the particles inside the avalanches show that the displacement / rotation field is chaotic. This is confirmed by the direct analysis of these fields (as in Fig. 6) which are not presented here for concision. Thus the failure is not localized, but rather of diffuse nature (however strain localizations have been found in experiments involving foundations).

Figure 10 compares the variations of the global second order work and the global kinetic energy versus the rotation angle. In this case the results show a large event between 15° and 20° and 2 smaller events for 3° and about 7°. These conclusions can be exhibited from both Figures 10-a and 10-b, what is showing a correlation between negative second order works and granular avalanches.

From a quantitative point of view, the results in Figures 8 and 10 are very different. In the first case, D^2W is varying between -0.5 10^{-3} J and +0.5 10^{-3} J, while in the second case the amplitude of variations are 4 times higher (from -2 10^{-3} J to +2 10^{-3} J). The highest value of the kinetic energy E_c is equal to 25 J in the first case and to 135 J in the second case (5 times higher). However in both cases the physical mechanism of failure is constituted by granular avalanches and the second order work seems to be a proper tool for the analysis.

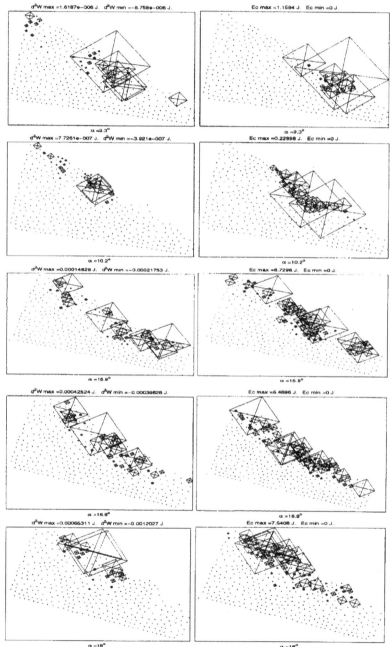

Figure 9. Negative values of the discrete second order work (on the left) and values of kinetic energy (on the right) for different rotation angles of the box.

76

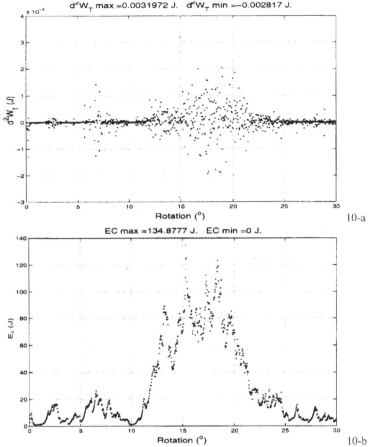

d²W_T max =0.0031972 J. d²W_T min =−0.002817 J.

10-a

EC max =134.8777 J.   EC min =0 J.

10-b

Figure 10. Variations of the global discrete second order work (at the top) and of the global kinetic energy (at the bottom) versus the rotation angle.

6 CONCLUSIONS

The objective of this paper was to revisit briefly the question of failure in geomaterials. The most usual definition of failure is its characterization by "limit stress" state i.e. by the vanishing values of the constitutive determinant what represents a plastic limit condition. The notion of flow rule comes as a direct consequence. Besides as recalled in the first part of this paper, these limit stress states are unstable according to Lyapunov's definition and to Hill's condition. In the first part also, according to the theory of non-associated elasto-plasticity and after the experiments, it has been exhibited a whole stress domain of instabilities, losses of uniqueness, bifurcations which are basically failure states.

Some of these failure states correspond to localized modes of failure with shear band, while others are associated to "diffuse" modes of failure, where the displacement field has a chaotic nature without localization pattern.

For an incrementally piecewise linear constitutive relation, the equations of the boundary of the unstable stress domain and the equations of the cone of unstable stress directions have been determined for axisymmetric conditions.

In the second part of the paper, "granular avalanches" (as called by the physicists of granular media) have been considered in the framework of discrete mechanics. It has been shown that Hill's condition of stability might be a pertinent indicator of failure for granular avalanches, because of its close link with the bursts of kinetic energy.

ACKNOWLEDGEMENTS

The French national contract ROMICO, the European project DIGA and the RNVO program are gratefully acknowledged for their supports.

REFERENCES

Bigoni D., Hueckel T. 1991. Uniqueness and localization-I. Associative and non-associative elastoplasticity , *Int. J. Solids Structures*, 28(2) : 197-213.

Nova R. 1994. Controllability of the incremental response of soil specimens subjected to arbitrary loading programmes, *J. Mech. behav. Mater.*, 5(2) : 193-201.

Vardoulakis I., Sulem J. 1995. Bifurcation Analysis in Geomechanics, Chapman and Hall.

Desrues J. 1984. La localisation de la déformation dans les milieux granulaires. Thèse d'état, Grenoble.

Darve F., Laouafa F. 2000. Instabilities in granular materials and application to landslides, *Mech. Cohes. Frict. Mater.*, 5(8) : 627-652

Darve F., Chau B. 1987. Constitutive instabilities in incrementally non-linear modelling. *In constitutive Laws for Engineering materials*, Desai C. S. ed. : 301-310.

Herrmann H.J., Hovi J-P., Luding S. 1998. Physics of dry granular media, Kluwer Academic Publ.

Bak P. 1996. How nature works. Springer Verlag publ.

Joer H., Lanier J., Fahey M. 1998. Deformation of granular materials due to rotation of principles axes. *Geotechnique*. 48(5) : 605-619.

Moreau J-J. 1994. Some numerical methods in multibody dynamics, application to granular materials, *Eur. J. Mech. A.*,13(4) : 93-114.

Lyapunov A.M. 1907. Problème général de la stabilité des mouvements. *Annales de la faculté des sciences de Toulouse*. 9 : 203-274.

Hill R. 1958. A general theory of uniqueness and stability in elastic-plastic solids, *J. Mech. and Phys. Solids*. 6 : 239-249.

Darve F., Labanieh S. 1982. Incremental constitutive law for sands and clays, simulations of monotonic and cyclic tests, *Int. J. Num. Anal. Meth. In Geomech.* 6 : 243-275.

Darve F., Laouafa F. 2001. Modelling of granular avalanches as material instabilities, *In Bifurcation and Localisation in Geomechanics*, Muehlhaus *et al.* Eds. Zwets and Zeitlinger publ. : 29-36.

Lanier J., Jean M. 1999. Experiments and numerical simulations with 2D-disks assembly, *Powder Technology, Spec. Issue on Numerical Simulations of particulate Media*, C. Thornton ed.

Cundall P.A., Strack O.D.L. 1979. A discrete numerical model for granular assemblies, *Geotechnique*, 29(1) : 47-65.

Bifurcations & Instabilities in Geomechanics, – Labuz & Drescher (eds.)
© 2003 Swets & Zeitlinger, Lisse, ISBN 90 5809 563 0

Fluctuations and Instabilities in Granular Materials

Junfei Geng, R. R. Hartley, D. Howell, and R. P. Behringer
Duke University, Durham, NC 27708

G. Reydellet and E. Clément
Université Pierre et Marie Curie, Paris 75231, France

ABSTRACT

We consider the properties of force chain networks in static or slowly deforming granular systems. We provide a brief description of the experimental method. We then present experimental determinations of the Green's function for static systems. That is, we determine the force response to a small locally applied force. We then consider the statistical properties of a slowly sheared granular system in a Couette geometry. All of these experiments are carried out in a quasi 2D geometry, as explained below.

INTRODUCTION

The goals of these studies are multiple. They include a determination of how forces are transmitted in granular materials through Green's function measurements, and characterizations of the slow shearing in dense materials.

First, regarding Green's function, the nature of force propagation has been a subject of intense recent debate (Nedderman, 1992, Liu et al. 1995, Bouchaud et al. 1995, Claudin et al. 1998, Goldenberg and Goldhirsch, 2002). A host of recent as well as older models are in competition, and it is crucial to provide clear experimental tests to distinguish among them or possibly to point to new directions. The differences among these models are substantial: there is not even agreement on the order of the appropriate differential equation (in the continuum limit). Older models (Nedderman, 1992) arising from the context of soil mechanics are well known in this group; for instance elasto-plastic models are described by elliptical PDE's below failure, and hyperbolic equations above. Coppersmith et al. (Liu et al. 1995) have proposed a model, based on a lattice construction, that is diffusive in character, and hence parabolic in the long wave length limit. Bouchaud et al. (Bouchaud et al. 1995, Claudin 1998) have proposed a model that is hyperbolic in the limit of a highly ordered system and that becomes elliptic as the amount of disorder increases. Most recently, Goldenberg and Goldhirsch (2002) have proposed newer elastic models. One approach to distinguish among these models is to carry out Green's function measurements, i.e. the determination of the force response to a small locally applied force (Reydellet and Clément, 2001, Geng et al. 2001, 2002). In theory, the point force should be infinitesimal. In practice, it must be small but finite, and the response should be linear in the applied force. It is in this regime that we operate.

A second goal is to characterize the statistical properties of dense granular systems as they undergo slow shear deformation (Howell et al. 1999). Here, there are several interesting questions.

Under steady shear, as in the Couette experiment used here, a dilated region, a shear band, forms that provides a transition between a solid and fluid-like regime. Shear bands are a dominant feature of many shear flows, yet their properties are not well understood. More generally, as the density of a sheared system changes, there is an overall transition from a dense to a more dilated state. The statistical and mean properties are of general interest, and provide insight into the nature of jamming (Liu and Nagel, 1998).

Finally, conventional models of sheared dense granular systems assume rate-invariance. In experiments described here, we show that rate dependence is not actually followed. Rather, the force network increases logarithmically in strength with increased shear rate.

Experimental Techniques

All the experiments described here are two-dimensional in nature. The particles used were either thin flat disks or thin flat particles with a pentagonal cross section, and all were cut from a photoelastic material. By using a polariscope, consisting of crossed circular polarizers, it is possible to determine forces at the particle scale. Here, we use a calibration (Geng et al. 2001, 2002) that takes photoelastic image data and provides a measure, G, that is proportional to the local pressure within an accuracy of about 10%. A sketch of the basic technique is given in Figure 1.

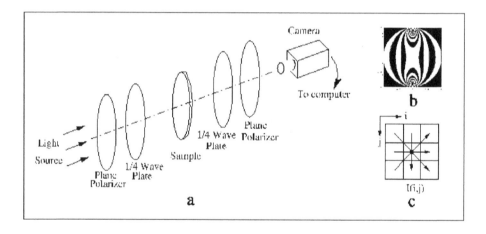

Figure 1. Sketch of a polariscope. The basic operation involves two crossed circular polarizers. In part b, we show the photoelastic image expected from a single particle subject to opposing forces. The density of light and dark fringes increases with increasing applied force and we use this fact to generate an empirical calibration of applied pressure vs. the gradient square of the photoelastic image (calculated as suggested by part c) averaged over the particle size. We denote this quantity as G^2.

In the Green's function experiments (Geng et al. 2001, 2002) we created packings, such as that sketched in Figure2. Typically, these packings were oriented nearly vertically, by stacking particles within a frame that is backed by a smooth, powder lubricated glass plate. This plate is tilted by about 2 degrees from vertical so that the packing was stable, and so that any friction with the plate was negligible. We then observed the change in the force chain network after the application of a small (typically 50g) force that was applied to one grain. Figure 2 shows a photoelastic image before the application of a point force (a), after the application of a point force (b), and the change in the force network (c,d) for two different realizations. Before the

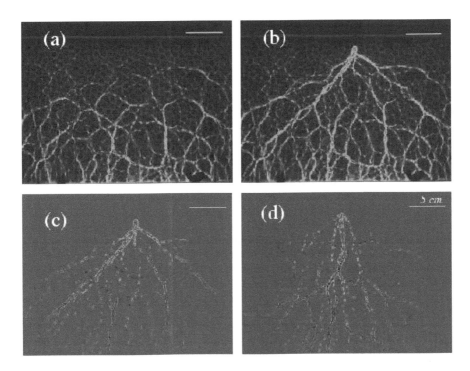

Figure 2. Photo elastic images showing a) the response before applying a point force; the force network is then due only to gravity; b) after the application of a point force; and c,d) the difference images (digital subtraction of two frames after and before the application of the point force).

application of the point force the image shows stress chains that are due only to hydrostatic head. The difference image ununderscores the complexity of the response: any two such realizations are in fact different.

Due to the complexity and variety of these responses, we proceed by obtaining an ensemble of results.

That is, we repeat the nominally identical experiment multiple times, in this case 50 times, in order to obtain both the mean behavior and the expected variability. This latter point seems particularly important if one is have a good handle on predictability.

In the Couette shear experiments, we place the photoelastic particles on a smooth horizontal flat surface, bounded from the inside by a rough wheel that provides the shearing, and on the outside by an equally rough ring (see below for a sketch). The whole is placed in a polariscope for photoelastic measurements. We imaged a key-stone segment of the experiment in real time (30 frames per second), and processed these data to obtain particle velocities, spins and stresses at the particle scale. It was then possible to provide detailed statistical data on these quantities.

In a related set of experiments, we repetitively compressed a sample of particles that was confined on three sides by rigid rectangular walls and on the fourth by an oscillating piston. We provide more detail schematics of the experiments in the sections below.

GREEN'S FUCNTION MEASUREMENTS

An extensive set of measurements has been carried out (Geng et al. 2002,2002) to determine the Green's function under a variety of conditions. These are detailed elsewhere, and here, we focus on a small but important subset of these measurements. These measurements focus on the role of order and disorder. Here, disorder has at least two origins. One of these is simply that amount of structural ordering of the grains. The other involves the fact that, even for a perfectly ordered packing of particles, the frictional forces are typically disordered. That is, the frictional force at a contact may fall anywhere in the range $|F_f| \leq \mu |F_N|$. The actual values are constrained by force balance, but also reflect the microscopic and generally unknown details of how the grains were placed.

We contrast the mean responses for several different packings in the following figures. First, we show the mean response for a spatially ordered hexagonal packing of monodisperse disks, Figure 3a. This is shown in greyscale and in a more quantitative form in Figure 4. The grey-scale part in particular gives a good overall picture: the force is, on average, carried most strongly along two primary lattice directions. In addition, the peaks associated with the strong propagation directions broaden with distance from the applied force. This is consistent with the model of Bouchaud et al (1995) and Claudin et al. (1998). However, it is also consistent with an anisotropic elastic model (Goldenberg and Goldhirsch, 2002). At this time, it is not possible to distinguish between these two cases.

If we increase the amount of disorder in the system by changing to packings of bidisperse disks, Figures 3b and 4b, we find that the mean response can still maintain a two-peak structure for modest spatial disorder. However, with a modest amount of disorder, the response becomes a single broad peak. By changing to a packing of pentagons, Figures 3c and 4c, a packing with no long-range positional ordering, we find a response that is a single broad peak that broadens steadily with depth. A plot of the width of this peak, Figure 4c inset, varies linearly with depth. Note that this is consistent with an elastic response. That is, for an infinite elastic half plane subject to a normal point force, all stress components show a single peak response with widths that vary linearly with distance from the source. This response is not consistent with a variation as the 1/2 power with depth that would be expected from the q-model.

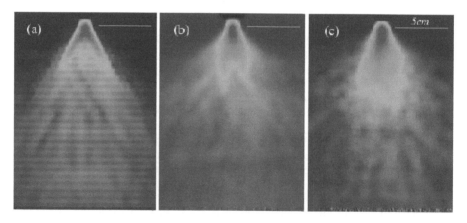

Figure 3. Greyscale representation of the statistically averaged point force response function for a) regular hexagonal packings of monodisperse disks, b) somewhat disordered packings of bidisperse disks, and c) packings of pentagonal particles.

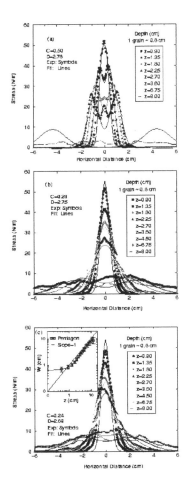

Figure 4. Quantitative representations of the greyscale data of Figure 3. Here, the ensemble mean force response is shown along cuts for various depths into the packing. Here, the top of the packing is at z = 0. The various parts of the figure are for a) regular hexagonal packings of disks, b) packings of bidisperse disks, and c) packings of pentagons.

Finally, we can increase the order in the packing by changing to rectangular packings, and by decreasing the friction coefficient of the particles. Rectangular packings, such as that shown in Figure 5, have a minimal number of contacts for stability. (To maintain this packing in a stable state, we used a template consisting of a set of grooves at the base.) Also, the range of allowed friction forces for a given normal contact force is reduced as the friction coefficient decreases. We show responses for two different rectangular packings in Figure 6. The first case is for relatively frictional particles. The second case corresponds to a packing of the same particles in the same geometry, but the particles were coated with Teflon tape. These packings show two sharp peaks. The peaks from the lower friction particles broaden more slowly than those for the higher friction particles. As for the hexagonal packings, it is not possible at this time to discriminate between the Bouchaud et al. model (essentially hyperbolic) and the Goldenberg-Goldhirsch anisotropic elastic picture, and a new generation of experiments is needed.

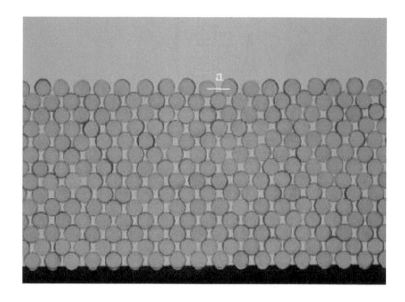

Figure 5. An image of a rectangular monodisperse disk packing.

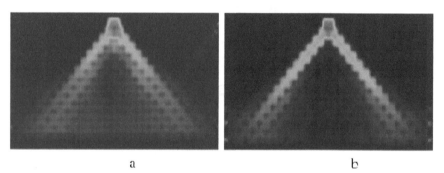

| a | b |

Figure 6. Mean force response for regular rectangular packings of monodisperse disks for relatively high friction particles, a) and relatively low friction particles, b).

STATISTICAL PROPERTIES OF DENSE SLOWLY SHEARED SYSTEMS

We turn now to the statistical properties of granular systems undergoing shear in a Couette geometry. We are particularly interested in the transition that occurs as the mean density of the system, as expressed by the packing fraction γ, is varied through a critical value such that stress chains just propagate through the system. As noted above, this system exhibits a shear band in the vicinity of the inner shearing wheel. Near the shearing wheel, the density is reduced. In addition, the velocity profile varies roughly exponentially with distance from the shearing wheel, as shown in Figure 7. Note also the interesting oscillations in the particle spin (particle rotation rate) near the shearing wheel. In the next few figures, we document the fact that as gamma is varied, system properties show features of a critical-like transition. These include a vanishing order parameter (the mean stress) critical slowing down, and a growing length scale (associated with the length of stress chains). For instance, in Figure 8, we show the mean stress vs. γ-γ$_c$ for a bidisperse system of disks. These data are consistent with a power law with exponent 4.0. In Figure 9, we show the velocity profile scaled by the distance from the transition point, V* = v/ε vs. a weakly rescaled distance from the shearing wheel, r* = r/(1 + A ε), where epsilon = (γ-γ$_c$)/γ$_c$, and A is a constant. Note that there is good collapse of the data for a range of densities. Finally, we have estimated the distribution of force chains as a function of the packing fraction and from these distributions we have computed the mean stress chain length vs. γ. These data, Figure 10, show that the mean stress chain length grows (although not necessarily diverges) as ε vanishes.

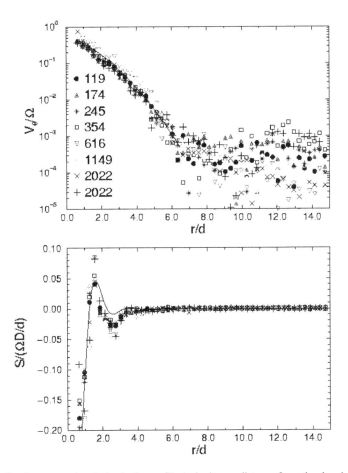

Figure 7. Data for the mean azimuthal velocity profile (velocity vs. distance from the shearing wheel), top and for the mean particle rotation rate or spin. These data have been scaled appropriately by the particle diameter d and or the rotation rate, Ω. Also, D is the diameter of the shearing wheel.

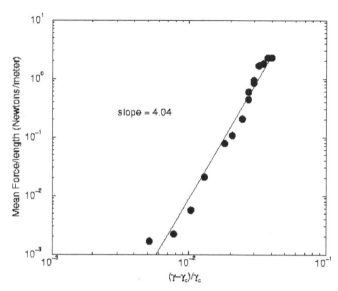

Figure 8. Mean stress vs. the reduced packing fraction, $(\gamma-\gamma_c)/\gamma_c$.

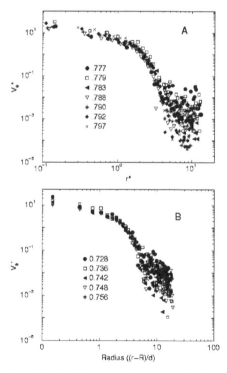

Figure 9. Rescaled velocity profiles for disks top, and pentagons, bottom.

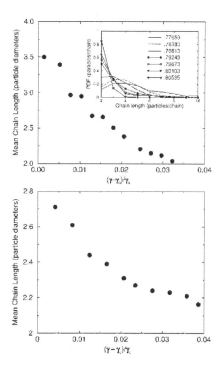

Figure 10. Mean force chain lengths vs. reduced packing fraction near the critical transition for top, disks, and bottom, pentagons. Inset top: distributions of force chain lengths for several packing fractions (for disks).

In the last part of this work, we describe experiments to examine the issue of rate-independence for the slow shear of dense materials. In fact, the present experiments show that there is logarithmic rate-dependence for sheared systems, a rate-dependence that does not appear in cyclically compressed collections of the same types of particles. As with all the above experiments, we used photoelastic techniques. Typically, we imaged roughly 10% of segment of the shear cell. Images were acquired a 15 frames per second, and were processed on the fly to obtain the mean stress. A set of typical set of stress time series for three different shear rates, Ω, Figure 11, give an intuitive sense of the effect of increasing shearing rate. Lower rates yield stress time series of lower amplitude that are also more intermittent. We show the mean of a number of these time series vs. shear rate Ω in Figure 12. These were obtained for various densities and shear rates spanning a bit over three orders of magnitude. All these data are consistent with a logarithmic variation of stress with Ω, as shown by least squares fit lines. Note that the stress network strengthens as Ω grows.

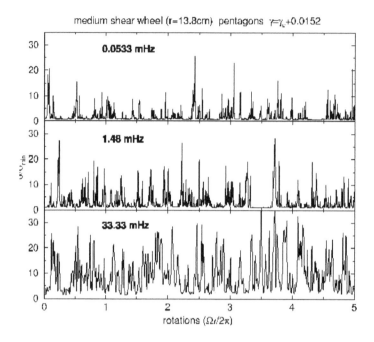

Figure 11. Integrated stress images vs. time, scaled by the rotation rate. Numbers indicate $\Omega/2\pi$.

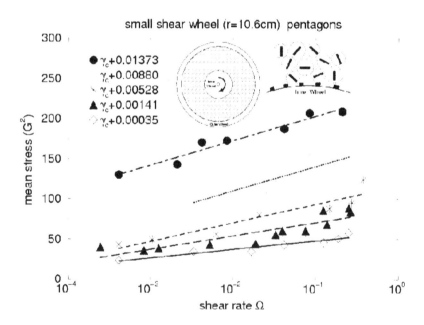

Figure 12. Mean stress (given by G^2) for several packing densities of pentagons. These densities are referenced to the critical value discussed above.

Also of interest are the statistical properties of the growth and relaxation of stress in the system. We show some of these in Figure 13. We define a growth or relaxation ("avalanche") event as any monotonic increase (decrease) of the mean stress in the system over time. Two properties that characterize these events include the stress magnitude, $\Delta\sigma$, and the angle θ through which the shearing wheel turns during the event.

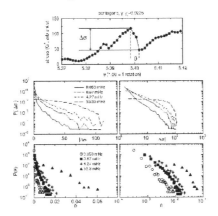

Figure 13. Statistical properties of avalanche events on semi-log and log scales.

If shearing is suddenly stopped, the stress decays over relatively long time scales, as in Figures 14 and 15. In fact, relaxation still typically occurs after 10s of hours. This slow relaxation suggests a possible origin for the strengthening of the force chain network with increasing rate: very

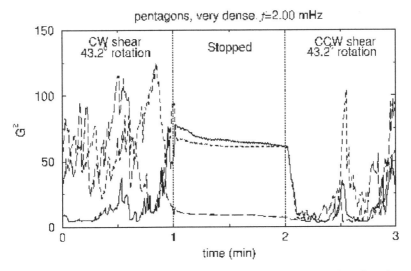

Figure 14. Fluctuations and relaxation after stopping or changing the direction of rotation.

slow relaxation of the network means that stress fluctuations do not decay before new fluctuations are generated.

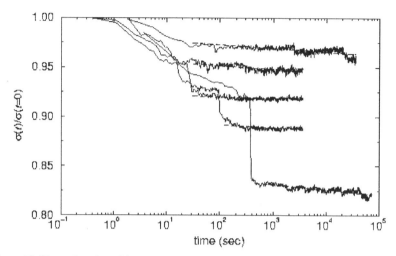

Figure 15. Slow relaxation of the stress network over very long time scales for several representative cases. Here, we have normalized by the stress at the starting time.

If particles are subjected to repetitive compression and released, no such rate dependence is observed, as seen in Figures 16 and 17. The inset of Figure 17 sketches the geometry of the piston

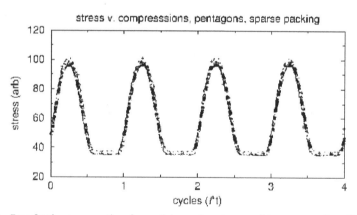

Figure 16. Data for the stress vs. time for particles undergoing repetitive compression. Over long times, these data collapse on top of each other.

experiment from which the data are drawn. The amplitude and frequencies of compression were chosen so that the range of piston velocities and the rates match those for the shear experiments. This experiment suggests that rate-dependence occurs when at least some of the inter-grain contacts are close to slipping.

90

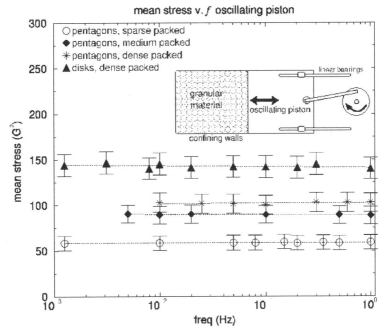

Figure 17. Data for the mean stress for various types of particles undergoing repetitive compression at the indicate frequency and an amplitude of 0.4 cm (particle sizes are about 0.7 cm).

CONCLUSIONS

We have explored force propagation and fluctuations in dense static or slowly deformed systems. For the static case, we have shown that order, whether due to geometric packing or frictional contacts, strongly influences the nature of force propagation. Highly ordered packings show propagation along principal lattice directions that broaden with distance from the applied force and with the amount of disorder. Highly disordered systems show a single peak response where the width of the peak grows linearly with the distance from the applied force, as in an elastic model.

In sheared Couette systems, there is a critical-like transition that occurs as the packing fraction is varied through a critical value γ_c of the packing fraction. This transition corresponds to the point at which stress chains vanish, and is likely related to the jamming transition. In Couette flow, there is logarithmic rate dependence of the stress network such that with increasing Ω, the network strengthens. This logarithmic increase with Ω, appears to be related to slow relaxation processes seen when shearing is suddenly stopped. Similar rate dependence is not present in the absence of shearing, i.e. under repetitive compression and relaxation. We note that previous studies (Nasuno et al. 1999, Ovarlez et al. 2001, Losert et al. 2000, Karner and Marone, 1998) have seen rate dependence in granular systems. However, these studies are unique in that they reveal changes within the force network.

REFERENCES

Bouchaud, J.-P. & Cates, M. E. & Wittmer, J. P. & Claudin, P. 1995, Stress Distribution in Granular Media and Nonlinear Wave Equation, *J. Phys. I, France* 5; 639-656.

Claudin, P. & Bouchaud, J.-P. & Cates, M. E. & Wittmer, 1998, Models of stress flutuations in granular media, *Phys. Rev. E*, 57: 4441-4457.

Geng, J. & Howell, D. & Longhi, E. & Behringer, R. P. & Reydellet, G. & Vanel, L. & Clément, E, & Luding, S. 2001, Footprints in sand: The response of a granular material to local perturbations, *Phy. Rev. Lett.* 87: 084302.

Goldenberg, C. & Goldhirsch, I. 2002, *Phys. Rev. Lett.* 89: 084302.

Howell, D. & Veje, C. & Behringer, R. P. 1999, Stress Fluctuations in a 2D Granular Couette Experiment: A Continuous Transition, *Phys. Rev. Lett.* 82: 5241-5244.

Karner, S. L. & Marone, C. 1998, The effect of shear load on frictional healing in simulated fault gouge, *Geophys. Res. Lett.* 25: 4561-4564.

Liu, C.-h. & Nagle, S. R. & Schecter, C. A. & Coppersmith, S. N. & Majumdar, S. & Narayan, O. & Witten, T. A. 1995, Force fluctuations in bead packs, *Science*, 269: 513.

Liu, A. L. & Nagel, S. R. 1998, Nonlinear Dynamics: Jamming is just not cool anymore, *Nature*, 396: 21-22.

Ovarlez, G. & Kolb, E. & Clément, E. 2001, Rheology of a confined granular material, *Phys. Rev. E* 64: 060302(R).

Nasuno, S. & Kudrolli, A. & Bak, A. & Gollub, J. 1998, Mechanisms for slow strengthening in granular materials, *Phys. Rev. E* 58: 2161-2171.

Nedderman, R. M. 1992, *Statics and Kinematics of Granular Materials*, Cambridge University Press.

Reydellet, G. & Clément, E. 2001, Green's Function Probe of a Static Granular Piling, *Phys. Rev. Lett.* 86: 3308-3311.

Bifurcations & Instabilities in Geomechanics, – Labuz & Drescher (eds.)
© 2003 Swets & Zeitlinger, Lisse, ISBN 90 5809 563 0

Avalanche Instability and Friction Mobilization : A Two-Phase Stability Limit

L. Staron, J.-P. Vilotte
Institut de Physique du Globe de Paris, France

F. Radjai
Laboratoire de Mécanique et Génie Civil de Montpellier, France

Keywords: Slope stability, frictional interactions, granular texture, numerical simulation

ABSTRACT: The evolution of 2D granular piles driven towards the stability limit by gravitational loading has been numerically investigated using the non smooth contact dynamic method. During the evolution, despite an intrinsic gradient loading effect, the texture can be classically analyzed in terms of a strong and a weak contact force networks with characteristic stress contributions. The system evolution is also characterized by intermittent dynamical events, driven by local contact slip instabilities, which lead to reorganize the granular packing and the texture. This can be described in term of an order parameter related to the density of critical contacts defined as the relative number of contacts where the friction is fully mobilized. A statistical analysis show spatio-temporal correlations of the critical contacts which are organized spatially as clusters. The clusters are localized domains of low pressure which can not sustain shear stress increments and appear as fluid (or plastic) bubbles. The mean size of the clusters exhibits a power-law divergence as the system approaches the stability limit which is reminiscent of a percolation process. These results, together with the analysis of the granular texture, suggest that granular material can be analyzed as a bi-phasic material and the destabilization process as second order transition.

1 INTRODUCTION

Understanding the behavior of granular materials is of major importance at least as a physical paradigm for a wide variety of natural and engineering applications. In geophysics for example, this has long been invoked for slope instabilities and rock debris avalanches. However, this behavior is far from being understood and the transition between a solid-like state, in equilibrium under shear stress loading, and a fluid-like state still remains a puzzling problem (Jaeger, 1996). Such a transition is of major importance in the understanding of the mobilization of a granular system under gravity loading, especially in the case of slope instabilities.

In the classical Coulomb's plasticity framework, the mechanical behavior of the granular material is simply characterized by an internal angle of friction φ defined as the maximum slope angle at which the granular material becomes unstable and start flowing (Nedderman,1992). Such an angle can be related to an effective friction coefficient μ_{eff}, $\mu_{eff} = \tan(\varphi)$. The stability limit of a slope thus occurs when the ratio of the tangential stress τ to the normal stress σ_n has reached μ_{eff}, the Coulomb plastic threshold.

Granular slope stability has been shown however to be more complex. In continuum mechanics non associative plasticity, for which plasticity and stability limit does not coincide even in the case of homogeneous loading, is todays advocated based a wide number of laboratory experiments. In physics, the hysteresis of slope instability has been experimentally emphasized with the evidence of a metastable state for slope angles in between the static and the dynamic angles of repose. Clearly, the mechanical behavior of a granular slope is the result of the multiple contact interactions and the induced geometry of such discrete structure. However such a renormalization is far from being understood. The identification of the pertinent macroscopic variables and evolution laws which govern the response and the stability of granular materials remain an active subject of research. Numerical simulations at the grain scale provide a unique tool which allow a direct measurement of local variables describing the evolution of the discrete structure geometry, the kinematics and the local forces transmitted through the grain contacts during minimal experiments of granular slope destabilization.

In this paper, the destabilization of a granular pile in response to gravity solicitations is numerically analyzed. The system consists in a rectangular granular bed tilted at a constant rate up to the stability limit where a surface flow avalanche is triggered. In the first section, the numerical model and the experimental setup are briefly introduced. The macroscopic behavior of the granular system is then outlined in terms of macroscopic friction properties and stress evolution(section 3). The granular texture is analyzed as well as its relation with the macroscopic evolution (section 4). Then a detailed statistical local analysis of the friction mobilization is presented (section 5). The notion of critical contact is first introduced and allow to define an order parameter that describe the intermittent evolution of the system characterized by dynamical events related with local frictional instabilities. The spatial correlations of the critical contacts are analyzed as well as their relation with the local stress state. Finally based on the analysis of the critical contacts, the avalanche instability is discussed in term of a percolation process and in the context of a two-phase instability (section 6).

2 NUMERICAL SETUP

2.1 *Non Smooth Contact Dynamics*

The numerical experiments were performed using the Non Smooth Contact Dynamics (NSCD) method of Moreau (1988) and Jean(1994). We refer to these references for more details in this method. It is worth however to mention here that the method is based on an implicit solution of the contact conditions, Signorini and friction, without any regularization. This allows for an accurate resolution of the onset of relative sliding between grains. A simple Coulomb friction law is assumed between grains in contact. This friction law can be pictured as a graph, Figure 1, relating, for a given normal force f_n, the tangential force f_t to the relative sliding velocity. The tangential force at the contact is therefore restrained to be $f_t \in [-f_y, f_y]$, where $f_y = \mu f_n$ with μ the coefficient of friction between the grains. When the friction is only partially mobilized, i.e. $|f_t| < f_y$, no relative slip motion can occur. When $f|f_t| = f_y$, the friction is fully mobilized and relative slip can occur. Such a friction law is clearly non-smooth and coupled to the Signorini contact condition through the normal force. Both problems has to be solved implicitly using a splitted iterative strategy.

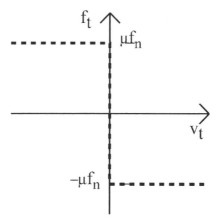

Figure 1: Contact friction law relating the friction force at contact f_t to the relative slip velocity v_t. The Coulomb friction threshold is given by the coefficient of friction μ and the normal force at contact f_n.

Figure 2: Initial state of a granular pile: the free surface is horizontal and nearly flat. The width of the bed is of $\sim 125D$, the thickness of $\sim 30D$.

2.2 The experiment

The system is made of two-dimensional circular particles, cohesion-less and perfectly rigid. The particle diameters are uniformly distributed in the range $[D_{min}, D_{max}]$ with $D_{min}/D_{max} = 2/3$. This slight polydispersity introduces a geometrical disorder in the packing that would otherwise likely exhibit a crystal-like organization. Collisions are assumed to be purely dissipative, an assumption which is reasonable for rock debris flows. However a systematic investigation of the influence of the coefficient of restitution has yet to be done but it is expected that for weak coefficient of restitution the results of this study will be found quite robust. In all the simulations presented here, the coefficient of friction between particles is set to $\mu = 0.5$. A systematic study on the influence of the friction coefficient has been performed and has shown that the main results are quite robust. It will not however be discussed here for sake of place but presented in a forthcoming paper.

The granular samples in this study are granular beds obtained by a random rain of 4000 particles within a initially horizontal rectangular box. The bed thickness is $\simeq 30D$, where D is the mean diameter of the particles. The initial free surface is nearly flat and horizontal (Fig 2). The granular bed are then continuously tilted under gravity loading up to the stability limit and the triggering of a surface flow avalanche. The rate of tilting is set to bet constant at $1°s^{-1}$. During each numerical experiment, the angle θ of the free surface slope evolves therefore from an initial value $\theta = 0$ to a limit angle $\theta = \theta_c$ defined as the angle at which a surface avalanche is triggered. Moreover, the analysis is based on 50 independent realizations.

3   MACROSCOPIC FRICTIONAL PROPERTIES

The distribution of the critical angle θ_c, at which the surface flow avalanche is triggered, is shown Fig 3 and has been calculated for 50 independent realizations. The distribution is centered with a mean value around $20.5°$. In the classical Coulomb analysis, θ_c can identified as the internal friction angle φ, and allows to define the effective coefficient of friction μ_{eff} for the granular bed :

$$\mu_{eff} = \tan(\theta_c) \simeq 0.37 \ .$$

The effective friction coefficient is found to be lower that the coefficient of friction at the contacts between the particles, and is a non trivial result from both the contacts geometry and the microscopic friction μ (Pöschel 1993).

The stress tensor σ, for a given volume V, is classically defines (Kruyt 1986) as :

$$\sigma_{ij} = \frac{1}{V} \sum_{\alpha \in N} f_i^\alpha l_j^\alpha, \tag{1}$$

where N is the total number of contacts in V, f_i^α is the i-component of the force transmitted at the contact α, and l_j^α the j- component of the vector joining the centers of mass of the two grains involved in the contact α. With that definition, the evolution of the ratio τ/σ_n can be drawn, Fig 4, as a function of the slope angle θ. Such an evolution exhibits linear increase from an almost zero initial value to a value of $\tau/\sigma_n \simeq 0.28$ when $\theta = \theta_c$ and an avalanche is triggered. Then as the avalanche evolves, a softening behavior is attested by the decrease of τ/σ_n. This type of material response is reminiscent of the behavior of dense sand samples under shear or compression loading (Darve 2000). As a result of the surface avalanche, the surface slope decreases from θ_c to a smaller value at which the system get stabilized again.

Regarding θ_c as the internal angle of friction of the packing, we should expect that at the stability limit:

$$\tau/\sigma_n(\theta_c) = \tan(\theta_c) = \mu_{eff}.$$

However from Fig 4, it is found that $\tau/\sigma_n(\theta_c) < \mu_{eff}$. This discrepancy with Coulomb's prediction is difficult to interpret and might be related to geometrical effects due to the geometry and finite size of the box. Indeed, the two vertical walls induces a gradient in pressure along a direction parallel to the free surface that should be taken into account when computing the mean averaged stress state.

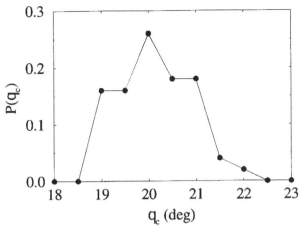

Figure 3: Distribution of the angles θ_c.

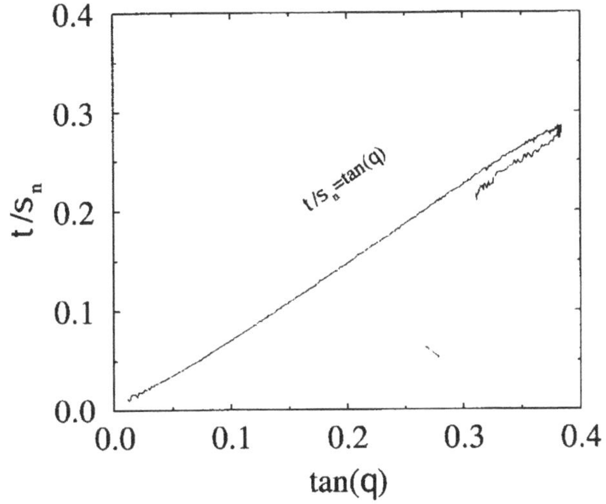

Figure 4: Evolution of the ratio τ/σ_n as a function of the slope angle θ.

4   THE GRANULAR TEXTURE

Based on the analysis of the distribution of the contact forces in shearing experiments, it has long been recognized that forces are transmitted along chains or contact force network. The contact force network can be splitted into a weak and a strong network according to the distribution of the force moduli (Radjai, 1996 ; Mueth, 1998). The strong network is composed of the contacts that transmit forces stronger that the mean force modulus in the system while the weak network only transmit forces weaker that this averaged force. Forces of the strong network are shown to be spatially correlated with a correlation length of the order of 5 to 12 D.

In contrast with shearing experiments, the gradient effect induced by the gravity loading has to be filtered out when renormalizing the forces with respect to the mean force. This is done by analyzing the forces within strips parallel to the free surface.

4.1 *The weak and strong contact networks*

From the analysis of the distribution of the contact forces moduli transmitted in granular packings, two "populations" of contacts (or contact sub-networks) can be distinguished regarding the mean force modulus (Mueth (1998), Radjai (1996)). Those contacts transmitting forces weaker than the mean force are said to belong to the weak contact network, whereas contacts transmitting forces greater than the mean belong to the strong contact network. Moreover, the strong contacts are spatially correlated and form "force chains" *e.g.* privileged transmission paths, which define a mesoscopic scale ξ roughly ranging from 5 to 12 D.

Analyzing the forces distribution in our samples, we obtain that 40% of the contacts are strong and form force chains of typical length $\xi = 8 - 10D$, as can be calculated from the force correlation fonction and can be seen from Fig 5. These characteristics of the texture remain mainly unchanged in the course of the evolution of the samples.

4.2 *Evolution of the partial stress tensors*

Using the stress tensor definition given by the equation 1, we can define the partial stress tensors σ^s and σ^w, where the subscripts s and w referred to the strong and the weak contact network respectively. The partial tensors are related to the global stress tensor through the relation:

$$\sigma = \rho_s \sigma^s + \rho_w \sigma^w, \tag{2}$$

where ρ_s and ρ_w are the relative proportion of strong and weak contacts respectively ($\rho_s \simeq 0.4$ and $\rho_w \simeq 0.6$).

For each of these stress tensors, the ratio of the tangential stress to the normal stress with respect to the slope, τ^s/σ_n^s and τ^w/σ_n^w can be computed. The evolution of both ratios as a function of the slope θ is reported Fig 6. For the strong network contribution, the stress ratio exhibits a regular increase with θ up to θ_c, whereas for the weak network contribution it remains quite constant and very low. The deviatoric loading is therefore almost screened out for the weak network and essentially taken into account within the strong network. The strong contact network behaves as a solid skeleton and is responsible for the mechanical strength of the granular pile. Such a result is in very good agreement with previous analysis of the texture in the context of shearing numerical experiments (Radjai, 1988) and stress out the heterogeneity induced by the discrete internal structure of granular materials.

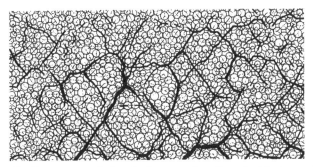

Figure 5: Force transmission in the granular piles: in black are drawn the force chains, the intensity of which is proportional to the lines width, whereas weak forces appears as thin gray lines.

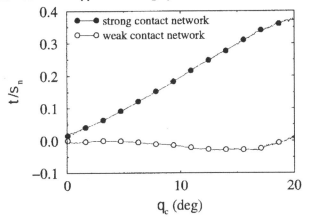

Figure 6: Evolution of the ratio τ/σ_n as a function of the slope angle θ in the strong contact network (filled circles) and in the weak contact network (empty circles).

5 MOBILIZATION OF FRICTION: THE CRITICAL CONTACTS

In order to get some insights on the destabilization, the analysis can be focused on the mobilization of the friction forces now at the contact scale. Based on the Coulomb Friction law (Fig 1), the contacts can be splitted into two subsets based on the value of the friction force f_t : either under the Coulomb threshold, $f_t < \mu f_n$, with no slip motion possible , either at the Coulomb threshold, $f_t = \mu f_n$, with possible slip motion. In the latter case, the contact can no more sustain an increment of the deviatoric load without a relative slip event between the particles. Such contacts are denoted here as *critical contacts*.

For a volume V of material, and for a given slope angle, the density of critical contacts ν is simply defined as the number of critical contacts N_c in that volume normalized by the total number of contacts of the volume. This characterize the mobilization of the friction within a given volume.

$$\nu(\theta, V) = \left\langle \frac{N_c}{N}(\theta) \right\rangle_V.$$ (3)

5.1 Intermittency and local instabilities

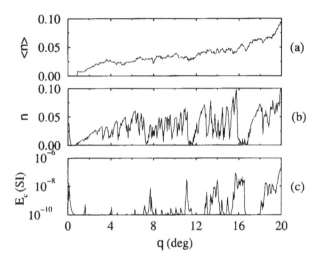

Figure 7: Evolution as a function of the slope θ of : (a) and (b) ν, (c) the kinetic energy. The curve (a) is obtained after averaging over 50 independent realizations while (b) and (c) are typical for a single realization.

The evolution of ν, evaluated for the whole system, as a function of the slope θ, is shown Fig 7(b) for a single realization. A stick-slip evolution of ν can be observed with loading phases, characterized by an increase of critical contacts, followed by intermittent relaxation events which induce a sudden loss of critical contacts. Such relaxation events corresponds to localized collective rearrangements of the geometrical configuration triggered by slip instability at critical contacts. An overall increase of ν can also be observed, from zero to $\simeq 0.08$ at the stability limit, which account for the progressive mobilization of the system. The relaxation events are dynamical events as observed in the evolution of the kinetic energy of Fig 7(c). The amplitude of the kinetic energy fluctuations, increases as the system approaches the stability limit, as well as the number of the critical contacts involved in the rearrangements. These relaxation events can be observed far before the slope has reached the critical value θ_c, and can be seen as precursors to the destabilization.

During the evolution ν tends to a limiting value of $\nu_c \simeq 0.8$. This suggests a saturation phenomenon and the pile reaches the stability limit when ν tends to a critical state ν_c at which a global dynamical reorganization occurs. This critical density ν_c has been shown to characterize the transition between static equilibrium and dynamical state (Staron et al., 2002) and the parameter $\phi = \nu/\nu_c$ can be considered as an order parameter.

5.2 Spatial correlations of the critical contacts

Spatial correlations of the critical contacts can be analyzed by computing the probability to find a critical contact at a distance r of a critical contact during the tilting evolution. The distribution functions $P_\theta(r)$ are therefore defined as :

$$P_\theta(r) = \frac{1}{\Delta\theta} \int_{\theta-\Delta\theta/2}^{\theta+\Delta\theta/2} \frac{d\theta}{N_c(\theta)} \sum_{\alpha=1}^{N_c(\theta)} \nu_\alpha(\theta,r), \qquad (4)$$

where $\Delta\theta = 2°$ is an interval window over which the function P_θ is averaged due to the intermittency, r is the radius of the neighborhood $v(r) = \pi r^2$, $N_c(\theta)$ is the number of critical contacts in the system for a slope angle θ, and $\nu_\alpha(\theta,r)$ is the local value of ν defined in the neighborhood $v(r)$ of the critical contact α.

The variations of the functions P_θ, with respect to the neighborhood radius r, are reported Fig 8, averaged over 50 independent realizations. These functions are decreasing functions of the neighborhood size r, which shows that the probability of finding a critical contact is always higher in the vicinity of a critical contact. For all the slope angles θ, the critical contacts tends therefore to form clusters. This behavior is highly reproducible from run to run. The characteristic length L of the decreasing functions P_θ can be seen as correlation length L for the critical contacts. From Fig 8, L can be estimated to be of the order of $8D$, close to the typical length ξ of force chains (section 4).

For all θ, the P_θ tends at large r toward the density value of critical contact ν evaluated over the whole system at θ tat we denote here $\nu_\infty(\theta)$. When plotting $P_\theta = (P_\theta(r)/\nu_\infty(\theta) - 1)$, as a function of r, in a log-log scale (Fig 9(a)), the relation:

$$\ln(P_\theta) = \gamma(\theta) + \eta(\theta)r$$

is shown to be verified as long as r remains inferior to $10D$. The functions P_θ can therefore be approximated as :

$$P_\theta(r) = \begin{cases} \nu_\infty(\theta)\,(1 + e^{\gamma}r^\eta) & \text{if } r \le 10D, \\ \nu_\infty(\theta) & \text{if } r > 10D, \end{cases} \qquad (5)$$

where the coefficients $\gamma(\theta)$ and $\eta(\theta)$ are affine functions of θ (Fig 9(b)).

When analyzing the evolution at the system scale (section 4.2), it has been proposed that the parameter $\phi = \nu/\nu_c$ could be regarded as a state parameter with $\phi = 1$ corresponding to the global stability limit. Locally, clusters of critical contacts for which $\nu = \nu_c$ are therefore of special interests. From the equation 5, the mean size of these clusters in the state ν_c can be deduced when replacing $P_\theta(r)$ by ν_c and rewriting the equality under the form:

$$r(\nu_c,\theta) = r_c(\theta) = \left[e^{-\gamma} \left(\frac{\nu_c}{\nu_\infty(\theta)} - 1 \right) \right]^{1/\eta}. \qquad (6)$$

We can compute the evolution of the mean size of the clusters of critical contacts in the state ν_c as a function of the slope angle θ. This evolution is reported in Fig 10. The size of these clusters remain almost constant as long as θ is less than $\sim 16°$. When this value is reached, r_c increases and finally diverges at the stability limit where the whole packing is in the state $\nu = \nu_c$. From the inset graph in Fig 10, the evolution of r_c can be defined as a power law $r_c(\theta) \propto (\theta_c - \theta)^{-\beta}$, where $\beta \simeq 0.45$. This power law divergence of r_c is reminiscent of a percolation process, involving two "phases".

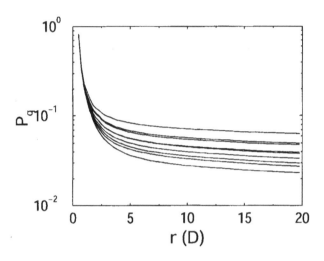

Figure 8: Variations of the functions P_θ with regard to r for different values of θ.

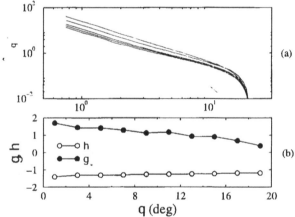

Figure 9: (a) Log-log representation of the functions $\mathcal{P}_\theta(r)$ (see text), (b) Evolution of the coefficients γ et η as a function of the slope θ.

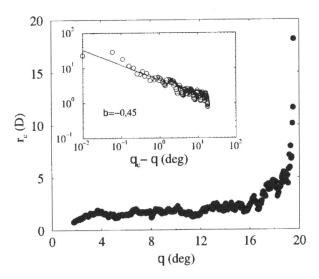

Figure 10: Evolution of the mean size r_c of the clusters of critical contacts in the state ν_c as a function of the slope angle θ. Inset graph: log-log representation of r_c as a function of $(\theta - \theta_c)$.

5.3 The local stress state

The stress state of these clusters of critical contacts can be analyzed. Following equation 1, we define the local stress tensor σ^α in the neighborhood $v(r)$ of a contact α:

$$\sigma^\alpha_{ij}(r) = \frac{1}{v(r)} \sum_{\alpha' \in v(r)} f_i^{\alpha'} l_j^{\alpha'}, \tag{7}$$

where α' denotes all contacts in the neighborhood $v(r)$ of α, $f_i^{\alpha'}$ is the i-component of the force transmitted at the contact α', and $l_j^{\alpha'}$ is the j-component of the vector joining the centers of mass of the two grains involved in the contact α'. Considering only the contacts α that are critical, and averaging over θ-windows of $\Delta\theta = 2°$ as previously done (equation 4, the mean stress tensor in a neighborhood $v(r)$ of the critical contacts for a slope angle θ is :

$$\sigma_\theta(r) = \frac{1}{\Delta\theta} \int_{\theta-\Delta\theta/2}^{\theta+\Delta\theta/2} d\theta \, \frac{1}{N_c(\theta)} \sum_{\alpha=1}^{N_c(\theta)} \sigma^\alpha(r). \tag{8}$$

The mean pressure and the mean deviatoric load of the clusters of critical contacts for a slope angle θ are therefore :

$$p_\theta(r) = \frac{1}{2}(\sigma_{\theta 1} + \sigma_{\theta 2})(r),$$

$$q_\theta(r) = \frac{1}{2}(\sigma_{\theta 1} - \sigma_{\theta 2})(r),$$

and

$$\mathcal{R}_\theta(r) = \frac{q_\theta}{p_\theta}(r) = \frac{\sigma_{\theta 1} - \sigma_{\theta 2}}{\sigma_{\theta 1} + \sigma_{\theta 2}}(r),$$

103

where $\sigma_{\theta 1}$ and $\sigma_{\theta 2}$ are the eigenvalues of the stress tensor $\boldsymbol{\sigma}_\theta$.

The variations of these quantities as a function of the size r of the neighborhood $v(r)$, and for the different values of θ, are reported in Fig 11. The clusters of critical contacts appear as areas of lower pressure and higher deviatoric loads, and the ratio of these two quantities $\mathcal{R}_\theta$ is the highest in the clusters of critical contacts. These clusters appears therefore as the result of a screening of the pressure within the system due to the heterogeneity in the stress transmission.

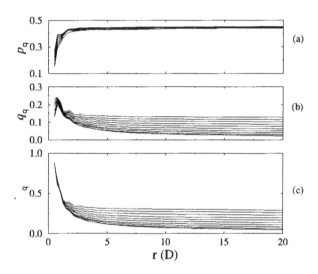

Figure 11: For different values of θ, evolution of the functions (a) p_θ, (b) q_θ and (c) $\mathcal{R}_\theta$ with regard to the radius r (see text for the definition of these functions).

6 THE CRITICAL CONTACTS AND THE TEXTURE: A SCENARIO FOR THE DESTABILIZATION PROCESS

We have studied the state of the granular packings during their evolution toward destabilization both in terms of granular texture (section 4) and in terms of the mobilization of microscopic friction (section 5). When studying the modulus of the forces transmitted through the critical contacts, we observe that more than 95% of the critical contacts belongs to the weak network (see Fig 12). The dynamical reorganizations triggered by the instability of the clusters of critical contacts take place in the weak contact network. The stability of the granular appears therefore as the result of interactions between the strong and the weak network rather than being triggered solely by the strong network.

Taking into account the granular texture and the evolution of the clusters of critical contacts, we can propose a mechanism for the destabilization of granular slopes. The packings are characterized by a strong contact network which plays the role of a solid skeleton and is responsible for the mechanical strength of the piles through force chains. While the packings are evolving toward the stability limit, critical contacts appears in the weak contact network, forming clusters. Clusters characterized by a critical state v_c can not sustain any shear increment and appear as plastic, or fluid, bubbles where local dynamical instabilities can occurred. During the tilting, as the slope angle θ increases toward θ_c, the mean size of the clusters of critical contacts in a plastic state $\nu = \nu_c$ increases and becomes of the order of the typical length ξ of the force chains. At this stage, the dynamical rearrangements triggered by the instability of theses clusters can destroy the stability of the force chains, and thus the overall stability of the packing. This is clearly what happens when the mean size of the clusters of critical contacts diverges as θ tends to the value θ_c.

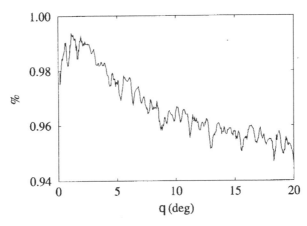

Figure 12: Proportion of critical contacts belonging to the weak contact network as a function of the slope angle θ.

7 CONCLUSION

The results we have presented in this paper stem from an accurate analysis of the state of granular packings evolving toward stability limit, and is an attempt to understand the local mechanisms acting during the slope destabilization. We have first characterized the effective friction properties of the granular matter. Than, the analysis of the granular texture has allowed us to identify a strong and a weak contact network exhibiting different mechanical response. The main part of this work is devoted to the study of the critical contacts, namely the contacts where the friction forces have reached the Coulomb threshold. We give evidence of intermittent dynamical instabilities related to the existence of clusters of critical contacts. This behavior reveals the existence of a critical density of critical contacts ν_c, so that the ratio $\phi = \nu / \nu_c$ can be seen as an order parameter, being zero in the initial state and reaching one at stability limit. From the analysis of the spatial distribution of the critical contacts, we show that they form clusters with a power law divergence of their mean size at the stability limit. This behavior is reminiscent of a percolation process. Moreover, we observe that the clusters of critical contacts are areas of lower pressure and that they belong to the weak contact network. The destabilization process can be seen as the result of the interactions between the weak contact and the strong contact networks.

This overall behavior is compatible with a bi-phasic picture of granular matter. The packing can be either "solid-like" when remaining at static equilibrium, or "liquid-like" when undergoing dynamical reorganizations. In this framework, the mobilization of the microscopic friction seems to be a relevant variable. This point of view has still to be compared to the theoretical work of Aranson (Aranson, 2001)) where a bi-phasic description of the granular matter is adopted using the relative value of the static and dynamical stresses as an order parameter. The power law divergence of the size of the clusters of critical contacts, the "liquid-like" part of the packing, suggest that slope destabilization can be seen as a second order phase transition. In this picture, the microscopic friction plays a role that we might compare to an effective temperature, as discussed by Degennes (Degennes, 1996).

REFERENCES

[1] Aranson I. S. & L. S. Tsimring, 2001, *Continuum description of avalanche in granular media*, Phys. Rev. E, 64, pR020301

[2] Darve F. & F. Laouafa, 2000, *Instabilities in granular media and application to landslides*, Mech. Cohes-Frict. Mater., 5, p627-52

[3] de Gennes, 1996, *Avalanche flows of granular mixtures*, J. Phys.

[4] Jaeger H. M., Nagel S. R. & Behringer R. P., 1996, *Granular solids, liquids, and gazes*, Reviews of Modern Physics, 68-4, p1259-73.

[5] Kruyt N. P. & L. Rothenburg, 1996, *Micromechanical definition of the strain Tensor for granular material*, J. App. Mech., 118, p706

[6] Jean M., 1994, *Frictional contacts in collections of rigid or deformable bodies: numerical simulation of geomaterial motion*, Elsevier Science Publishers B.V.

[7] Moreau J.-J., 1988, *Some Basics of Unilateral Dynamics*, unpublished.

[8] Mueth D. M., Jaeger H. M. & S. R. Nagel, 1998, *Force distribution in a granular medium*, Phys. Rev. E, 57 p3164-69.

[9] Nedderman R. M., 1992, *Statics and kinematics of granular material*, Cambridge University Press

[10] Pöschel T. & V. Buchholtz, 1993, *Static friction phenomena in granular materials: Coulomb Law vs. particle geometry*, Phys. Rev. Lett., 71, p3963

[11] Radjai F., Jean M., Moreau J.-J. & S. Roux, 1996, *Force Distributions in Dense Two-Dimensional Granular Systems*, Phys. Rev. Lett., 77, p274

[12] Radjai F., Wolf D., Jean M. & J. J. Moreau, 1998, *Bimodal character of stress transmission in granular packings*, Phys. Rev. Lett., 80, p61.

[13] Staron L., Vilotte J.-P & F. Radjai, 2002, *Pre-avalanche instabilities in a tilted granular pile*, Phys. Rev. Lett., 89, p204302-1

3. Modeling and localization

Bifurcations & Instabilities in Geomechanics, – Labuz & Drescher (eds.)
© 2003 Swets & Zeitlinger, Lisse, ISBN 90 5809 563 0

Shear bands in soil deformation processes

D. Leśniewska
Institute of Hydro-engineering, Gdańsk, Poland

Z. Mróz
Institute of Fundamental Technological Research, Warsaw, Poland

ABSTRACT: The present work extends the previous study (Leśniewska & Mróz 2000, 2001) and is related to the development of a characteristic pattern of shear bands and its role in the course of progressive deformation in soils. The recognition of physical mechanisms responsible for specific arrangements of shear band patterns is necessary to understand the post-failure soil behavior properly. The boundary value problem of granular material retained on one side by a flexible wall is used as an example. Some factors, supposed to have an influence on the patterning, are discussed. These are: the material softening, the weakening interaction at the soil-structure interface and the associated and non-associated flow rule. Both simple limit analysis and finite element calculations are employed. The conclusion is that the soil softening is not the necessary condition for strain localization, but it certainly enhances the whole process and makes it easier to observe. It also has some influence on the geometry of the shear band pattern. Some features of the pattern change also with the flow rule assumed .

1 INTRODUCTION

Strain localization phenomena were observed by numerous researchers in such materials as soils, rocks, metals or polymers. The localized modes usually develop from imperfections before the maximum stress value is reached by the material and grow due to progressive deformation within the localization zones, usually accompanied by material softening. Such structured deformation, containing shear band patterns, was observed for example in experimental studies by Desrues et. al (1996), Saada et. Al. (1994), Finno et al. (1997), Muir Wood (2001). The numerical analysis of shear band deformation was presented also by numerous authors, cf. Anand and Gu (2000), Ehlers and Volk (1997), Hicks and Mar (1994), Hicks (2000), Tejchman (2002).

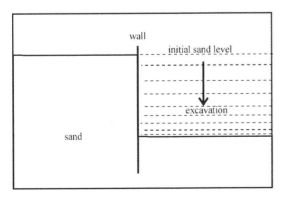

Figure 1. The outline of the tests on flexible cantilever walls.

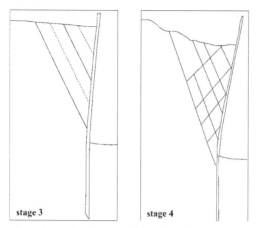

stage 3   stage 4

Figure 2.The final stages of shear band pattern evolution, taken from Milligan's radiographs (test GM 03).

Soil excavation from one side of a cantilever wall initially buried in sand constitutes a typical geotechnical problem frequently occurring in engineering practice. Small scale model tests on active soil pressure on cantilever walls were performed by Milligan (Milligan, 1974, Milligan and Bransby, 1976), Fig.1. The shear band pattern was detected by using X-rays. The details of the tests and the radiographs taken were recently reviewed by Leśniewska, 2000, and also analyzed theoretically by Leśniewska and Mróz (2000, 2001). Fig.2 presents an example of the shear band pattern, recorded on a radiograph during the test GM03, performed by Milligan on 14/25 (BS sieve) fraction of Leighton Buzzard sand. The test parameters are as follows: wall height h = 30 cm, wall flexibility number $2.97 \cdot 10^{-6}$, initial void ratio $e_o= 0.55$ (dense sand), smooth wall finish.

The existence of localized, multiple failure patterns, like those presented in Fig.2, is not usually taken into account in engineering calculations, specifying earth pressure distribution on a wall. Such calculations are usually based on the assumption of a single shear band, separating a rigidly moving soil block from the remaining soil. However, a sufficient experimental material is now available to support the assumption of failure mechanism composed of relatively widely spaced shear bands with progressive deformation evolving in consecutive bands and rigid motion of material blocks between the bands.

A simplified analysis of such failure mechanism was provided by Leśniewska & Mróz (2000, 2001), who considered limit equilibrium states associated with generation of consecutive bands. The soil was assumed as a rigid-plastic softening material with the cohesion or friction angle decreasing with the slip along active shear band. Due to elastic interaction with the supporting wall, the limit equilibrium occurs at the initiation of slip on the consecutive band with the termination of slip on the preceding band and the slip process occurring in the unequilibrated (dynamic) mode. Thus, the multiple shear banding was associated with the material softening within the bands. However, further insight into the problem indicates the possibility of shear band patterning also for a perfectly plastic material response. This is induced by softening mechanism of soil-wall interaction. In fact, the control parameter in this problem is the growing height of excavation on one side of the wall, thus inducing the transition of the cantilever wall support and the associated variation of wall compliance. This mode of contact softening will be discussed in more detail in the next section.

The numerical analysis of shear band pattern was performed by Mróz & Maciejewski (1994) for the cases of vertical wall punch or wedge penetration into the soil, using the kinematically admissible failure modes. The control parameter is then the progressive displacement of a rigid tool inducing plastic deformation in the active mode with the subsequent mode switching at discrete set of states on the deformation path. The elastic strains were neglected and a rigid-plastic response was analyzed by assuming material softening from the initial yield stress to its residual value.

Considering the specific dissipation functions D per unit volume and D^{Γ} per unit surface of the discontinuity surface Γ, the limit load factor was specified from the relation (1),

$$\int D(\dot{\varepsilon}, p)\, dV + \int D^{\Gamma}\left([v], p\right) d\Gamma - \lambda \int \underline{T}^{0} \cdot \underline{v}\, dS - \int \underline{f} \cdot \underline{v}\, dV = 0 \qquad (1)$$

where [v] denotes the velocity discontinuity on Γ, $\lambda \underline{T}^{0}$ is the traction on the tool interface, $\underline{f}$ is the body force vector and p is the hardening or softening parameter. The limit load multiplier λ can be determined directly from (1). As λ is then a function of changing configuration of a deforming material $\lambda = \lambda(q, p)$, where q denotes the control parameter inducing progressive deformation, the evolution of λ is schematically presented in Figure 3.

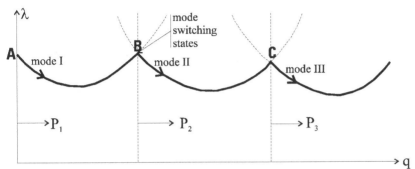

Figure 3. Progressive deformation of soil, accompanied by failure mode switching at B and C.

It is seen that after reaching the limit state at the initial configuration at A, the subsequent deformation occurs in mode 1 along the path AB, with initially decreasing and next increasing load factor λ. In fact, material softening in shear bands induced the decrease of λ, but the changing geometric configuration induces increase of λ. For the displacement-controlled deformation process the equilibrium path AB is terminated at B, when a new failure mode develops (mode II) and follows the equilibrium path BC with a new mode III, generating at C. Thus for a progressive deformation the mode switching occurs at B, C, ... with subsequent configuration –and state variable evolutions. The mode switching states are specified at B by the conditions (2):

$$\lambda_{I} = \lambda_{II} \qquad \frac{d\lambda_{I}}{dq} > \frac{d\lambda_{II}}{dq} \qquad (2)$$

Similar conditions occur at the consecutive mode switching states. A similar phenomenon is observed in considering post-buckling evolution of elastic structures, when the mode exchange may occur in the course of progressive deformation. When load control is applied, a dynamic transition between consecutive modes may occur and the equilibrium configuration depends on the kinetic energy acquired during the unstable portion of the deformation path.

The case of shear band patterning during progressive deformation of soil retained by a flexible wall can be treated similarly by introducing the generalized unequilibrated force, associated with the specific failure mode. The progressive excavation depth constitutes now the control parameter, monotonically increasing with time. In view of material softening or increase of the supporting wall flexibility in the course of increasing excavation depth, there are discrete states of limit equilibrium, specifying the generation and evolution of consecutive failure modes.

2 LIMIT EQUILIBRIUM MODEL

The simple limit equilibrium model, based on the Coulomb wedge analysis and published earlier, (Leśniewska and Mróz, 2000), was accepted in the present paper with certain modifications. In this model the equilibrium states associated with the deformed configuration are speci-

111

fied, but the transition to those states is assumed to follow through unequilibrated states. In reality, this transition can proceed as a dynamic process (requiring inertia forces to be introduced) or as a quasi-static process with additional dissipation within the moving wedge. However, the present paper follows a simplified approach based on the following assumptions: the failure mechanism consists of the consecutive sliding motion of soil wedges separated by several shear bands, the inclination of any activated shear band does not vary during the process of deformation, the soil wedge motion is affected at the same time by the increasing elastic soil-wall interaction and by the weakening of reaction force, both induced by the process of soil excavation. Outside the shear band the soil wedge is assumed to move as a rigid body. The thickness of shear bands is assumed as infinitesimal. In fact the shear band thickness is related to soil dilatancy and transverse strain gradient. However this effect is not considered at this stage of the analysis. Also, the effect of the second family of shear bands is not accounted for.

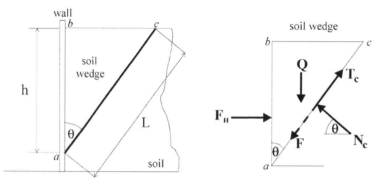

Figure 4. Equilibrium of forces in case of soil wedge motion.

The Coulomb wedge analysis is usually used to calculate earth pressures acting on a vertical retaining wall by considering the limit equilibrium of a soil wedge cut off from the soil mass by a slip plane ac (Fig.4). F means there the resultant, unequilibrated force, responsible for the potential wedge movement, N_c the normal soil reaction (total value), T_c – friction component of soil reaction (total value), Q – weight of the soil wedge and F_H – the wall reaction, assumed as horizontal for the sake of simplicity. The state of limit equilibrium occurs when all forces acting on a soil wedge are in balance (the resultant F of these forces equals zero). In such a case the soil friction is fully mobilised along the plane ac. It is possible, however, to consider the more general case of unbalanced forces, causing the rigid soil block to move (only downward "active" wedge movement is considered). Such a situation is illustrated in Fig.4 if F > 0. Projecting all forces acting on a wedge along the orientation of the shear plane, the value of a resultant unequilibrated force F can be specified. This force is conjugate to the sliding velocity and provides an extra work rate transformed into kinetic energy of the wedge motion. The source of the unbalanced force F lies here in the weakening of the soil-wall system due to soil excavation and in the soil-wall interaction compliance modulus.

It is also possible to consider wedge equilibrium not in the limit state when soil friction is not fully mobilised and the wedge cannot move. Referring to limit equilibrium when F = 0, we shall consider unequilibrated motion when F > 0 and equilibrium below the limit state when F < 0.

The Coulomb wedge analysis allows the critical height of a vertical cohesive cut to be determined, thus providing the maximum height which does not need an external support. According to this approach the critical height of a non-cohesive soil is equal to zero. This is the ideal case, however, when the presence of any supporting wall is neglected. In reality there is always some support from the wall. If this support is taken into account, it is possible to obtain a non-zero critical height for a cohesionless soil.

Milligan's radiographs show that the first shear band always appears at some distance from the top of the wall. This phenomenon can be explained in a simple way by the critical height concept (Leśniewska & Mróz, 2000).

The value of the unequilibrated force F in general case is given by formula (3), Leśniewska & Mróz, 2000).

$$F = \frac{1}{\cos\theta}\left\{\frac{1}{2}\frac{\gamma h^2}{\tan(\theta+\phi)+\cot\theta} - \frac{F_H(\theta,h,u)}{\cot(\theta+\phi)+\tan\theta} - ch\right\} \quad (3)$$

The soil unit weight, angle of internal friction and cohesion are denoted by γ, ϕ, and c respectively. It is assumed that the wall acts on the soil wedge through a horizontal force F_H and the friction of the soil-wall interface is neglected. The horizontal force F_H depends on θ and on the wedge displacement. It is clear that, depending on parameter values occurring in (3), one can consider two cases: $F \leq 0$, when the soil wedge is in equilibrium and $F > 0$, when the soil wedge moves downward along the shear band ac, Fig.4. Formula (3) represents a general case of the model, valid also for softening soil, when internal friction angle ϕ changes its value during deformation (Leśniewska and Mróz, 2000). For constant ϕ (no soil softening), c =0 and critical inclination of the soil wedges formula (3) can be simplified to the following form:

$$F = \frac{1}{4\cos\theta_c}\left[\gamma h^2 \tan\theta_c - 2F_H(\theta_c,h,z)\cot\theta_c\right] \quad (4)$$

where h means the height of the soil wedge, z the depth of excavation and $\theta_c = \pi/4-\phi/2$. The standard limit analysis refers to the forces applied to a soil body with rigid constraints. However, in the present case we need to account for elastic compliance of the retaining wall or compliance of soil near the contact plane. To preserve model simplicity, it is assumed that the soil-wall interaction is modeled by a set of elastic springs of uniform stiffness at the interface. In fact, the interaction stiffness increases toward the constrained wall end, but this effect is not considered. It is understood that the elastic properties of the soil-wall contact can be controlled by both the soil and wall elastic modulus. The assumed simplification, with uniformly distributed contact stiffness, allows for analytical treatment of the problem and clarifies the essential features of shear bands. In the previous version of the model (Leśniewska and Mróz, 2000) a uniform stress distribution p_0 was applied to the wall as a boundary condition. The origin of this distribution was not discussed. It was assumed that the distribution does not change during excavation. This assumption was justified by the fact, that mainly the influence of soil softening on shear banding in soils was investigated. Now we are trying to check, whether the shear banding without soil softening can exist. One of the possible ways is to take into account, that p_0 is an equivalent of the support given to the wall by a soil, which is not yet excavated (Fig.5). According to this assumption, the horizontal force F_H acting on a wall is given by the formula (5):

$$F_H = p_0(h-z) + E^w h\sin\theta = P(h, z) + E^w h\sin\theta \quad (5)$$

where p_0 denotes the boundary stress value, z denotes the depth of excavation, h the height of a soil wedge and u is the wedge displacement along the shear band. Relation (5) expresses in fact the weakening of the soil-wall system, combined with its strengthening due to elastic response of the soil-wall contact. The process of loosing and then recovering the equilibrium by a soil wedge is presented in Figure 5. To analyse the process, one has to detect whether any shear band can appear behind the assumed retaining wall. The maximum possible depth of excavation for Milligan's dredged tests (Milligan, 1974) is equal to the total wall height H $(0 < h < H)$. An active soil wedge can appear behind the wall at some excavation depth z_1, which can be related to the critical height h_{cr} calculated for the assumed initial wall pressure acting on a soil mass (Fig.5a). An active soil wedge can appear only if the critical height is smaller than the total wall height H. The critical height will be used later in this paper as a reference value h_{ref}. The critical height, according to its definition, has to be determined from the condition $F = 0$ at the initial rest configuration u = 0. It is understood that any soil movement may occur only along a plane defining the "critical soil wedge", the height of which is equal to h_{cr} and whose inclination to the vertical is equal to θ_c. The following expression for h_{cr} was obtained:

$$h_{cr}(u=0) = h_{ref} = \frac{2p_0}{\gamma}\cot\theta_c\tan(\theta_c+\phi) = \frac{2p_0}{\gamma}\cot^2\theta_c \quad (6)$$

113

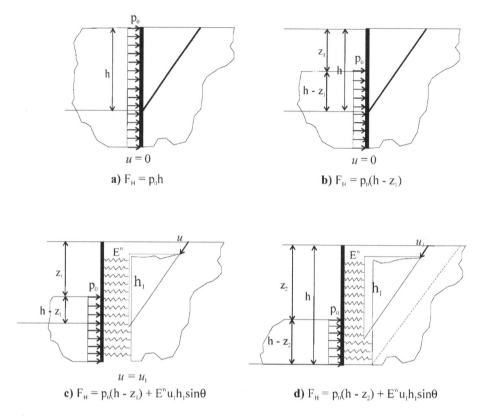

a) $F_H = p_0 h$

b) $F_H = p_0(h - z_1)$

c) $F_H = p_0(h - z_1) + E''u_1 h_1 sin\theta$

d) $F_H = p_0(h - z_2) + E''u_1 h_1 sin\theta$

Figure 5. Subsequent active soil wedges, appearing during soil excavation.

The procedure of determining locations of subsequent shear bands is illustrated in Fig.5. The starting point is the situation presented in Fig.5a. The depth of excavation $z = 0$, there is no wedge displacement ($u = 0$), full support is provided to the wall ($F_H = p_0 h$). The corresponding critical height of the soil wedge is given by (6).

Fig.5b shows the first stage of excavation: $z = z_1$, $F_H = p_0(h - z_1)$. We are looking for the height h of the soil wedge, which fulfils, for a given depth of excavation, the condition of limit equilibrium where F is given by (4). Combining (4) and (5), we obtain the following equation for h:

$$\gamma h^2 \tan\theta_c - 2p_0(h - z_1)\cot\theta_c = 0 \tag{7}$$

It is clear, that the value of h calculated from this equation has to be positive and not greater than h_{ref}, given by (6), thus the height of the first soil wedge, being in limit equilibrium, is equal to:

$$h = h_1 = \frac{p_0 \cot^2\theta_c}{\gamma}\left[1 + \sqrt{1 - \frac{2\gamma z_1}{p_0}\tan^2\theta_c}\right] \tag{8}$$

or, taking (6) into account:

$$h_1 = \frac{1}{2}h_{ref}\left[1 + \sqrt{1 - \frac{4z_1}{h_{ref}}}\right], \qquad 0 \le z_1 \le \frac{1}{4}h_{ref} \tag{9}$$

The smallest value of h_1 is obtained when $z_1 = h_{ref}/4$. Once the limit equilibrium is reached by the critical soil wedge, its motion is activated due to continuing excavation. The wedge displacement u can increase as long as $F \geq 0$. This stage of the process is presented in Fig.5c.

It was already pointed out that in the actual tests (Milligan, 1974), instead of one soil wedge moving along a single shear band there exist multiple wedges, appearing within a soil mass one after another. The condition necessary for the first wedge to appear was explained in the previous sections. The starting point for further discussion, regarding multiple wedges or multiple shear bands, is an existence of an already activated first soil wedge, which underwent some displacement and reached an equilibrium state. It is obvious that once the origin of the second wedge is explained, the occurrence of subsequent wedges follows the same rule. All the formulae derived for the first pair of shear bands are valid for any number of them.

According to experimental observation, any subsequent slip line is practically parallel to the previous ones. Hence, it is assumed that the new shear band inclination is the same as that of the first band.

The same equations of equilibrium are valid for the analysis of the second soil wedge (Fig.5d), provided the expression (10) is used for the horizontal force F_H:

$$F_H = p_0(h-z_2) + E^w h_1 u_1 \sin \theta_c \tag{10}$$

For this value of horizontal force a new critical height can be found for a cohesionless soil from the expression (4), defining the location of the second active shear band:

$$h_2 = \frac{p_0 \cot^2 \theta_c}{\gamma} \left[1 + \sqrt{1 - \frac{2\gamma}{p_0^2 \cot^2 \theta_c} \left(p_0 z_2 - E^w u_1 h_1 \sin \theta_c \right)} \right] \tag{11}$$

If (6) is used with (11), h_2 equals:

$$h_2 = \frac{1}{2} h_{ref} \left[1 + \sqrt{1 - \frac{4z_2}{h_{ref}} + \frac{4E^w u_1 \sin \theta_c}{p_0} \frac{h_1}{h_{ref}}} \right] \tag{12}$$

Having the location h_2, we can analyze the second wedge movement, which can last until a new state of equilibrium is reached. In the same way the subsequent locations of shear bands can be specified. Relations (9) and (12) prove that, in certain conditions, there can exist multiple critical soil wedges (and thus multiple shear bands) even if soil softening is not taken into account. It means that soil softening may not be a necessary condition for strain localization, however it certainly enhances the whole process and makes it more observable. It also has an influence on the structure of shear band pattern, as it was shown previously (Leśniewska and Mróz, 2000). We will try to support the above findings in the second part of this paper, using the finite element method solution.

3 FINITE ELEMENT ANALYSIS

The plane strain problem of soil excavation from one side of a cantilever wall, initially buried in sand, will now be analyzed by the finite element method. PLAXIS finite element code has been used to perform all computations, as it offers high order elements (15-nodes triangles, Fig.6). An elastic-perfectly plastic Mohr-Coulomb model, involving five material parameters, i.e. E and v for soil elasticity, ϕ and c for soil plasticity and ψ as an angle of dilatancy, was used in view of its simplicity (Plaxis Manual, 1998). The strain and strain rates are decomposed into elastic and plastic parts:

$$\underset{\sim}{\varepsilon} = \underset{\sim}{\varepsilon}^e + \underset{\sim}{\varepsilon}^p, \quad \underset{\sim}{\dot{\varepsilon}} = \underset{\sim}{\dot{\varepsilon}}^e + \underset{\sim}{\dot{\varepsilon}}^p \tag{13}$$

The Hooke's law is used to relate the stress rate to the elastic strain rate:

$$\underset{\sim}{\dot{\sigma}} = \underset{\sim}{D}^e \underset{\sim}{\dot{\varepsilon}}^e = \underset{\sim}{D}^e \left(\underset{\sim}{\dot{\varepsilon}} - \underset{\sim}{\dot{\varepsilon}}^p \right) \tag{14}$$

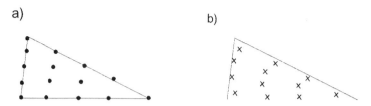

Figure 6. Finite element used for calculations– a) element nodes, b) stress points

The plastic potential function is introduced, along with the Coulomb-Mohr yield condition. The plastic strain rates are expressed as:

$$\dot{\underline{\varepsilon}}^p = \dot{\lambda}\frac{\partial g}{\partial \underline{\sigma}'}, \quad \dot{\lambda} > 0, \quad f = 0, \quad \dot{\lambda}f = 0 \tag{15}$$

where $\dot{\lambda} > 0$ is the plastic multiplier.

To study influence of different boundary conditions put at the toe of a flexible wall, two types of calculations were performed: with shorter wall, embedded only in sand, (h = 4.2 m) and with longer wall, fixed at the problem boundary (h = 5 m).

One of the purposes of the present paper is to show, that it is indeed possible to obtain full patterns of shear bands using classical elastic-perfectly plastic Mohr-Coulomb soil model with the associated flow rule. The effect of soil dilation on the process of shear band pattern forming was also studied by employing non-associated, non-dilative elastic-plastic soil model ($\psi = 0$). The flexible wall was modeled as a linear elastic beam (Plaxis Manual, 1998). A set of material parameters, used in both types of calculations, is listed in Tables 1 and 2.

Table 1. Soil parameters for non-associated and associated Mohr-Coulomb model

Nr	Type	γ_{dry} [kN/m^3]	γ_{wet} [kN/m^3]	v [-]	E_{ref} [kN/m^2]	c_{ref} [kN/m^2]	ϕ [$^\circ$]	ψ [$^\circ$]	$R_{interface}$ [-]
1	non-associated	17	20	0.3	13000	1	31	0	1
2	associated	17	20	0.3	13000	1	31	31	1

Table 2. Wall parameters

Number	Wall height [m]	Identification	Type	EA [kN/m]	EI [kNm^2/m]	v [-]
1	4.2	sheet-pile	elastic	6500000	10000	0
2	5	sheet-pile	elastic	6500000	10000	0

Two "rigid" (having the same strength as surrounding soil) interfaces were assumed along both sides of the wall to model soil-wall contact (R=1, Table 1). Standard boundary conditions (bottom boundary fixed, left and right boundary free to move in vertical direction) were used.

The forth order, 15-node triangular elements, available in PLAXIS, were chosen for a more accuracy in calculations, and because it was earlier proven, that this type of elements is capable of predicting soil collapse (Sloan and Randolph, 1982, Vermeer and van Langen, 1989).

A 15-node triangular element consists of 15 nodes and 12 stress points, as indicated in Fig.6. During a finite element calculation displacements are calculated at the nodes and stresses at individual Gaussian integration points (stress points). More detailed characteristics of 15-node triangle were presented by Sloan and Randolph, 1982.

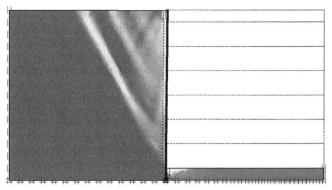

Figure 7. Incremental shear strain localization. Associated Coulomb-Mohr model assumed for soil (parameters set 2, Table 1). Wall 5m high (parameter set 2, Table 2)

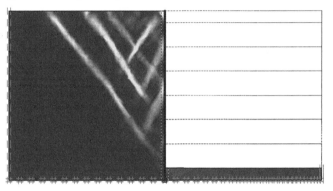

Figure 8. Incremental shear strain localization. Non-associated Coulomb-Mohr model assumed for soil (parameters set 1, Table 1). Wall 5m high (parameter set 2, Table 2)

All calculations were performed using professional 7.2 version of PLAXIS finite element code (Plaxis Manual, 1998). Mesh of triangle 15-nodes elements was employed. Five unstructured meshes, having different global coarseness, were used to perform the convergence study. All calculations, except those for convergence study, were performed on very fine mesh (1162 elements).

Small strain approach was used in a number of calculations. If larger deformations were expected, an updated Lagrange formulation was employed, available in PLAXIS. All calculations were perfomed with tolerance smaller or equal 3%. Soil excavation was modeled using PLAXIS "staged construction" option.

The example of finite element calculations is presented in Figures 7 and 8. They represent the isolines of incremental shear strains, obtained at the end of excavation and present final localization pattern, consisting of the two families of almost straight, almost parallel lines, clearly resembling the one recorded by Milligan in his tests (Fig.2).

Shear strain increments were chosen to present localization, because they are related to actual deformed mesh and due to it make the clearest picture, not influenced by the cumulating of previous deformation stages. It was also shown by Cramer et.al., 1998, that the distribution of the calculated error measures, used for adaptive mesh refinement, is related to plastic strain increments. It is understood, than, that they are good indicators of localization.

It is also possible to detect the same pattern, examining volume strain increments, however in case of non-associated flow rule the pattern looks very faint, in contrary to the associated flow rule case, like it should be expected. Cumulative measures of deformation, like total shear strains and total volume strains, also show traces of localization. It is obvious, however, that in this case, due to significant geometry changes during excavation, localization pattern is less dis-

117

tinct and more diffused. It is probably, because cumulative displacements represent values averaged over all stages of deformation. The pattern apparently preserves the locations of maximum deformations, however.

Figures 7 and 8 also show the influence of soil dilatancy on the shear band pattern. They represent two extreme cases: full dilation introduced by the associated flow rule and total lack of dilation given by $\psi = 0$ condition.

It is clear, that lack of dilation makes localization patterns easier to observe. It does not mean, however, that dilation suppresses localization – it only makes it more difficult to observe. The angles, shear bands make with vertical and horizontal, are also different in both cases. More work is necessary to prepare the material enabling direct comparison between calculated and experimental patterns. Looking at Fig.2 however, one has the impression, that at least the angles shear bands make in reality with vertical and horizontal are closer to the associate flow rule case.

4 CONCLUSIONS

The simple model presented in this paper constitutes an extension of the classical Coulomb wedge analysis and a modification of the model presented earlier (Leśniewska and Mróz, 2000). It has shown that the conclusions led previously were incomplete – the soil softening may not be the only necessary condition for strain localization, however it certainly enhances the whole process and makes it more observable. Soil softening has also an influence on the structure of shear band pattern, like it was shown previously.

Finite element calculations for classical elasto-plastic Coulomb-Mohr material also showed the existence of the pattern of strain localizations, thus supporting the opinion, that soil softening is not the only factor, leading to shear banding in soils. Further parametric studies are necessary to recognize more widely the mechanisms of strain localization and detailed structure of shear band patterns formed. It has to be also explained, why shear band patterns form in certain conditions. We regard the present approach, employing high order finite elements, as the most reliable to capture the shear band pattern.

REFERENCES

Anand, L., Gu, C. 2000. Granular materials: constitutive equations and strain localisation, Journal of the Mechanics and Physics of Solids, 48, 1701-1733

Cramer, H., Rudolph, G., Steinl, G., Wunderlich, W. 1998. A hierarchical adaptive finite element strategy for elastic-plastic problems, Computers and Structures, 73, 61–72.

Desrues, J., Chambon, R., Mokni, M., Mazerolle, F. 1996. Void ratio evolution inside shear bands in triaxial sand specimens studied by computed tomography, Geotechnique, 46, No. 3, 529–546.

Ehlers, W., Volk, W. 1997. On shear band localisation phenomena of liquid-saturated granular elastoplastic porous solid materials accounting for fluid viscosity and micropolar solid rotations, Mechanics of Cohesive-frictional Materials, Vol.2, 301-320

Finno, R.J., Harris, W.W., Mooney, M.A., Viggiani, G. 1997. Shear bands in plane strain compression of loose sand, Geotechnique, 47, No. 1, 149–165.

Hicks, M., Mar, A. 1994. Mesh adaptivity for geomaterials using a double-hardening constitutive law, Computer Methods and Advances in Geomechanics, Siriwardane and Zaman (eds), Balkema, Rotterdam

Hicks, M. 2000. Coupled computations for an elastic-perfectly plastic soil using adaptive mesh refinement, Int. J. Numer. Anal. Meth. Geomech., 24, 453–476.

Leśniewska, D. 2000. Analysis of shear band pattern formation in soil, Institute of Hydro-engineering of the Polish Academy of Sciences, Gdańsk.

Leśniewska, D., Mróz, Z. 2000. Limit equilibrium approach to study evolution of shear band systems in soils, Geotechnique 50, No. 5, 521–536.

Leśniewska, D., Mróz, Z. 2001. Study of evolution of shear band systems in sand retained by flexible wall, Int. J. Numer. Anal. Meth. Geomech., 25, 909-932

Milligan, G.W.E. 1974. The Behaviour of Rigid and Flexible Retaining Walls in Sand, PhD Thesis, University of Cambridge.

Milligan, G.W.E., Bransby, P.L. 1976. Combined active and passive rotational failure of a retaining wall in sand, Geotechnique, 26, No. 3, 473–494.

Mróz, Z., Maciejewski, J. 1994. Post-critical response of soils and shear band evolution, Proceedings of the Third International Workshop on Localisation and Bifurcation Theory for Soils and Rocks, Aussois, France, (Chambon, Desrues, Vardoulakis (eds)), Balkema, 19-32

Muir Wood, D. 2001. Some observations of volumetric instabilities in soils, IUTAM Symposium on Material Instabilities and the Effect of Microstructure, Austin, Texas

Plaxis Manual. 1998. Balkema, Rotterdam, Brookfield.

Saada, A.S., Bianchini, G.F., Liang, L. 1994. Cracks, bifurcation and shear bands propagation in saturated clays, Geotechnique, 44, No.1, 35-64

Sloan, S.W., Randolph, M.F. 1982. Numerical prediction of collapse loads using finite element methods, Int. Jnl Num. Anal. Meth. Geomechanics, Vol. 6, 47-76

Steinmann, P., Betsch, P. 2000. A localisation capturing FE-interface based on regularized strong discontinuities at large inelastic strains, International Journal of Solid and Structures 37, 4061-4082

Tejchman, J. 2002. Patterns of shear zones in granular bodies within a polar hypoplastic continuum, Acta Mechanica 155, 71-94

Vermeer, P.A., van Langen, H. 1989. Soil collapse computations with finite elements, Ingenieur-Archiv 59, No. 3, 221-236

Bifurcations & Instabilities in Geomechanics, – Labuz & Drescher (eds.)
© 2003 Swets & Zeitlinger, Lisse, ISBN 90 5809 563 0

Modeling of localization in granular materials: Effect of porosity and particle size

B. Muhunthan, O. Alhattamleh, and H.M. Zbib
Washington State University, Pullman, USA

ABSTRACT

The flow stress in the Mohr-Coulomb plastic constitutive equation is modified with a higher order gradient term of the effective plastic strain as a means of modeling the effect of inhomogeneous deformation in granular materials. The gradient constitutive model has been incorporated into a finite element code and used to simulate biaxial shear tests on dry sand. It is shown that the higher order gradient has a significant influence on the material behavior; especially when instabilities such as shear band take place. The gradient term acts as a stabilizer in the post localization regime where material exhibits strain softening. Granular material consists of discrete particles and associated voids. Therefore, a higher order gradient constitutive model must explicitly account for its gradients in terms of the heterogeneous granular microstructure. It is apparent that for the physical problem under consideration the main physical length-scale is the particle size. Furthermore, we argue that there are two main physical mechanisms that seem to control flow, hardening/softening and localization, namely, particle-particle contact (magnitude and heterogeneous directional distribution) and porosity (magnitude and heterogeneous spatial distribution). Therefore, we propose a continuum model that accounts for these mechanisms and associated length scales.

INTRODUCTION

Classical continuum models for elastoplastic deformation of materials are usually formulated for homogenously deforming media. Thus, they do not include internal length scales. These scales that may define the dimension of the underlying substructure, become relevant when deformation is inherently heterogeneous and local strain gradients may become intense, consequently, altering the material behavior. When the deformation fields are highly heterogeneous, phenomenological constitutive equations that do not consider the effect of high gradients fail to predict deformation patterns that may persist and eventually dominate the deformation mechanism such as shear bands.

Various methods in which length scales have been introduced to form non-local type continuum models are available in the literature. These methods include micropolar theories in which rotational degrees of freedom are added to the conventional translational degrees of freedom (Vardoulakis (1989), de Borst (1992), and Fleck and Hutchinson (1993)). Many studies have shown that a bifurcation analysis coupled with a micropolar theory makes it possible to predict the thickness of shear band as well as its direction. Micropolar theories, however, result in non-symmetric stiffness matrix and cause problems in finite element analysis.

Integral and gradient type theories forms another class of alternative methods to introduce length scales. In the integral type models (Bazant and Gambarova (1984)) the evolution of certain internal variables are expressed by means of integral equations. Mindlin (1964) introduced gradients into linear elasticity models to the form of non-local constitutive models. The motivation, significance and implications of introducing gradients into plasticity models has been highlighted in the works of Aifantis and co-workers (Aifantis (1987), Zbib and Aifantis (1988), Vardoulakis (1996)).

The strain gradient theory is an appropriate way to mathematically and physically model the effect of heterogeneity of plastic flow within a continuum mechanics approach. This, however, furnishes a new and interesting problem regarding the physical interpretation of the added material parameters "the gradient coefficients." Although the works of Zbib and Aifantis (1988-1989) provides a means for indirectly measuring these coefficients, no rigorous analysis has been performed which relates these coefficients to the size and scale of the underlying microstructure.

In this work a gradient type plasticity model of the form proposed by Zbib and Aifantis (1988) has been implemented in ABAQUS. It has been used to simulate the formation of shear bands and localization in a plane strain test on sand, showing the importance of including gradients terms in order to properly predict the post localization behavior. Next, the strain gradient method is extended to a physically based micro mechanical model that accounts for the inhomogeneity of plastic flow that results from heterogeneous microstructure, including porosity and grain morphologies.

MATHEMATICAL PRELIMINARIES

The deformation gradient F_{ij} of a continuous media undergoing a smooth deformation process is given by the relation:

$$F_{ij} = \frac{\partial \chi_i(X_j, t)}{\partial X_j} \tag{1}$$

where X_j is the position vector of a typical particle in the reference undeformed configuration and χ_i is its corresponding position vector in the current configuration. It is assumed that the body undergoes a large deformation process such that the resulting stretching is partly elastic and partly plastic. The velocity gradient L_{ij} is given as $L_{ij} = \dot{F}_{ik} F_{kj}^{-1}$. It can be split into two parts; symmetric and skew-symmetric. The symmetric part represents the pure stretching tensor, D_{ij}, and the skew symmetric part represents the spin rate tensor, W_{ij}:

$$D_{ij} = \frac{1}{2}(L_{ij} + L_{ij}^T); \quad W_{ij} = \frac{1}{2}(L_{ij} - L_{ij}^T) \tag{2}$$

The stretching rate D_{ij} can be decomposed as:

$$D_{ij} = D_{ij}^e + D_{ij}^p \tag{3}$$

where D_{ij}^e and D_{ij}^p are the elastic and plastic parts respectively. Similarly the spin tensor is written as:

$$W_{ij} = \omega_{ij} + W_{ij}^p \tag{4}$$

where ω_{ij} is the spin of microstructure and W_{ij}^p is the plastic spin.

Next we recall the equation of equilibrium for quasi static loading conditions:

$$\sigma_{ij,j} = 0 \tag{5}$$

where σ_{ij} is the Cauchy stress tensor.

For isotropic elastic solids the rate form of Hooke's law is adopted for the elastic stretching part D_{ij}^e:

$$\overset{o}{\sigma}_{ij} = C_{ijkl}^e D_{kl}^e \; ; \; C_{ijkl}^e = G(\delta_{ik}\delta_{jl} + \delta_{il}\delta_{jk} + \frac{2v}{1-2v}\delta_{ij}\delta_{kl}) \tag{6}$$

where C_{ijkl}^e is the elasticity tensor, G is the shear modulus, v is the Poisson's ratio. $\overset{o}{\sigma}_{ij}$ is the corotational rate of Cauchy stress tensor defined with respect to frame rotating with the material:

$$\overset{o}{\sigma}_{ij} = \dot{\sigma}_{ij} - \omega_{ik}\sigma_{kj} + \sigma_{ik}\omega_{kj} \tag{9}$$

D_{ij} can be split into a volumetric strain (D), and a deviatoric strain rate d_{ij} as:

$$D = D_{mm} \quad ; \quad d_{ij} = D_{ij} - \frac{1}{3}D\delta_{ij} \tag{10}$$

Similarly the components of the stress may be split into hydrostatic (p) and deviatoric (S_{ij}) parts:

$$p = \tfrac{1}{3}\sigma_{kk} \quad ; \quad S_{ij} = \sigma_{ij} - p\delta_{ij} \tag{11}$$

PLASTICITY MODEL

The yield function of the granular material is assumed to follow the Mohr-Coulomb criterion:

$$f = q - \kappa \tag{12}$$

where κ is the flow stress and q is the effective shear stress, defined as:

$$q = \sqrt{\tfrac{3}{2}S_{ij}S_{ij}} \tag{13}$$

Note that in conventional plasticity theories κ is assumed equal to $p\mu$ where μ is the coefficient of friction.

The plastic stretching d_{ij}^p is defined by means of a potential function Q as:

$$d_{ij}^p = \dot{\gamma}\frac{\partial Q}{\partial \sigma_{ij}} \tag{14}$$

where $\dot{\gamma}^p \left(=\sqrt{\tfrac{2}{3}d_{ij}^p d_{ij}^p}\right)$ is the effective plastic strain rate. The function for Q is assumed to be:

$$Q = q + p\beta(\gamma^p) \tag{15}$$

where $\beta(\gamma^p)$ is the dilatancy coefficient relating volumetric plastic strain to effective plastic shear strain as:

$$D^p = \beta(\gamma^p)\dot{\gamma}^p \tag{16}$$

Substituting Eq.(15) into Eq.(14) and combining with Eq.(16) leads to:

$$D_{ij}^p = (\frac{3S_{ij}}{2q} + \frac{\beta}{3}\delta_{ij})\dot{\gamma}^p \tag{17}$$

123

GRADIENT PLASTICITY MODEL

The gradient modification of the conventional plasticity here involves the modification of the flow stress κ (Eq. 12) as (Zbib and Aifantis, 1989):

$$\kappa = |p| \cdot \mu(\gamma^p) + c_1 \nabla^2 \gamma^p \; ; \quad \gamma^p = \int \dot{\gamma}^p \, dt \tag{18}$$

where c_1 is a gradient related coefficient. Note that in the above form higher order gradients are not included as their contribution was found to be minimal. Furthermore, other gradient modifications can be adopted as discussed in detail by Zbib and Aifantis (1989). With the incorporation of the gradient term the yield function becomes:

$$f = q - |p|\mu(\gamma^p) - c_1 \nabla^2 \gamma^p \tag{19}$$

Upon utilizing the yield and consistency conditions:

$$f = 0 ; \quad \dot{f} = 0 \tag{20}$$

the rate of plastic stretching could be evaluated as:

$$\dot{\gamma}^p = \frac{\dfrac{3 S_{ij} \dot{\sigma}_{ij}}{2q} - \dfrac{\dot{p}}{3} \mu \delta_{ij} - c_1 \nabla^2 \dot{\gamma}^p}{p \cdot h_t + c_1' \cdot \nabla^2 \gamma^p} \tag{21}$$

where $h_t = \dfrac{\partial \mu}{\partial \gamma^p}$, is the strain hardening/softening modulus and $c_1' = \dfrac{dc_1}{d\gamma^p}$, substituting Eq. (21) into Eq. (17) and combining it with Eqs. (3) and (6) leads to:

$$\overset{\circ}{\sigma}_{ij} = C_{ijkl} D_{kl} + C_1 S_{ij} + C_2 \delta_{ij} \tag{22}$$

where

$$C_{ijkl} = G[C^e_{ijkl} - C^p_{ijkl}] \tag{22a}$$

$$C_1 = \frac{c_1 \nabla^2 \dot{\gamma}^p}{\overline{H} q}, \; C_2 = C_1 = \frac{c_1 \alpha \beta \nabla^2 \dot{\gamma}^p}{\overline{H}} \tag{22b}$$

$$\overline{H} = 1 + |p|\frac{h_t}{G} + \alpha \beta \mu \tag{22c}$$

$$\alpha = \frac{K}{G} = \frac{1}{1 - 2v} \tag{22d}$$

Then, the resultant plastic stiffness tensor has the following form.

$$C^p_{ijkl} = \frac{1}{\overline{H}} (\frac{3 S_{ik}}{2q} + \alpha \beta \delta_{ik})(\frac{3 S_{jl}}{2q} + \alpha \beta \delta_{jl}) \tag{22e}$$

EVOLUTION OF FRICTION AND DILATANCY

The mobilized friction coefficient μ is assumed to be a function of the effective plastic strain as:

$$\mu(\gamma^p) = \mu_{cv} + x_1 \left(\gamma^p + \frac{\mu_0 - \mu_{cv}}{x_1} \right) exp(-x_2 \gamma^p) \tag{23}$$

where μ_{cv} is the internal friction at constant volume, μ_0 is the initial internal friction, and x_1, x_2 are fitting parameters. We note in passing that similar models relating μ_{cv} to μ have been proposed by other researchers (Balendran and Nemat-Nasser, 1993, Anand and Gu 2000).

Following the work of Taylor (1948) the dilatancy is expressed as (see also Vardoulakis 1996):

$$\beta(\gamma^P) = \mu(\gamma^P) - \mu_{cv} \qquad (24)$$

SIMULATION OF SHEAR BAND INITIATION

The gradient plasticity model has been implemented into the ABAQUS (2002) standard Finite Element code as a special user material subroutine. The model has been used to study the shear band initiation of some biaxial shear tests on dry sand reported by Han and Drescher (1993).

The biaxial apparatus used a prismatic specimen of 40 mm width, 80 mm length, and 140 mm height. The specimen was enclosed between two rigid walls 80mm apart and placed on a platen, which rested on a linear bearing. The linear bearing provided kinematic freedom for the formation of shear bands with the lower specimen portion sliding horizontally. The apparatus was placed inside a pressure chamber and the specimen was subjected to a confining pressure and kinematically or statically controlled axial load.

Biaxial tests were performed on coarse, poorly graded Ottawa sand with rounded particles of mean grain diameter $D_{50} = 0.72$ mm. Homogenous dense specimens, with an initial porosity of $n_0 = 0.32$-0.33, were prepared. All tests were performed with displacement controlled axial loading. Additional details of the test device and measurements of relevant parameters are provided in Han and Drescher (1993).

The biaxial test has been modeled here using plane strain finite element analysis. The discretized mesh arrangement is as shown in Fig. 1(a). All analyses were carried out using four-node plane strain element with reduced integration, CPE4R. The lower boundary of the mesh is assumed rigid. In the first phase, a confining pressure was applied to consolidate the specimen homogeneously. Thereafter, axial compression was applied by increasing the vertical displacement on the top of the specimen.

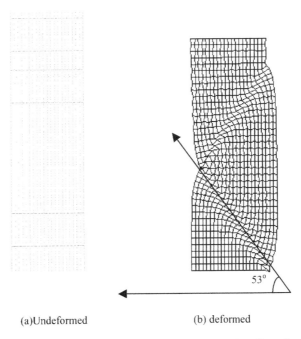

(a)Undeformed   (b) deformed

Fig. 1: 20x40 CPE4R elements. (a) undeformed configuration (b) deformed configuration showing the shear band inclination angle at 53°.

Based on the characteristics of the sand and the test conditions, the following material constants were used in the gradient Mohr Coulomb model:

Stiffness modulus	$E=200MN/m^2$
Poisson ratio	$v=0.104$
Initial friction	$\mu_o=0.100$
Constant Volume friction	$\mu_{cv}=0.613$
	$x_1=3.00$
	$x_2=-55.00$

Fig. 1(b) shows the shape of the deformed mesh with localization and shear band. In order to trigger the shear band, reduced values of parameters (x_1 & x_2) were assigned to one of the elements in the center of the FE mesh. Notice that since the lower boundary is constrained in all directions the shear band can also initiate from its corners. The inclination of the resulting shear band was measured to be $53°$ relative to horizontal axes (Fig. 1(b)). The experimentally measured angle was found to vary between $55°$ to $59°$ (Han and Drescher 1993).

INTERNAL LENGTH SCALE AND NUMERICAL SHEAR BAND THICKNESS

In order to give a physical attribute to the gradient coefficients, a material length scale was defined as:

$$l = \sqrt{\frac{c_1}{E}} \qquad (25)$$

where c_1 and E are as defined before. The thickness of the shear band is controlled by the length scale coefficient, l. Depending on the mesh size a different value of l was needed to obtain a consistent thickness of the shear band. For example, the measured shear band thickness was found to be 10.25mm for the model when the number of elements is 611 with $l = 4.33\mu m$, 800 with $l = 3.873\mu m$ and 1400 with $l = 2.872\mu m$. This shows that in order to avoid mesh-size dependency, it is important to include the effect of internal length scale by means of the gradient term.

Fig. 2 illustrates the effect of using strain gradient on the distribution of equivalent plastic strain within the specimen. It can be seen that the use of c_1 defuses the concentration of plastic strains and results in a consistent width of the shear band.

Fig.3 shows the effect of using the strain gradient term on the force displacement curve. All analyses were carried out at the same confining pressure of 200kPa. It can be seen that c_1 has a direct influence on the determining the shape of the curve.

The shape of the load displacement curve predicted by the analysis is dependent on the mesh size. However, as shown in Fig. 4 with the use of an appropriate gradient coefficient dependent on the mesh size, it is possible to predict a unique shape of the load displacement curve. Finer mesh required a finer length scale. Moreover the coefficient turns out to be approximately inversely proportional to the element area.

The analyses have shown the added flexibility offered by including higher order strain gradients to simulate strain localization in shear tests. One of the main drawbacks of such models, however, lies in the selection of the coefficients of the gradient terms such as c_1. The heterogeneous microstructure and its evolution with shear control these terms. Therefore, a micro mechanical model that accounts for the inhomogeneity of plastic flow resulting from a heterogeneous microstructure is needed. Such a model with appropriate averaging procedures may give the sought dependence of the flow stress on internal length scales and higher order strain gradients

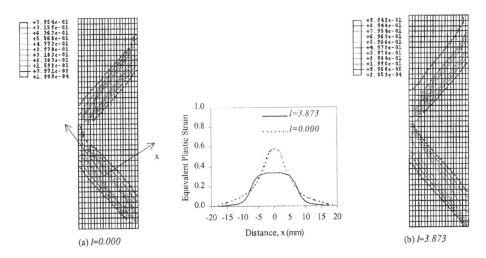

(a) *l=0.000*

(b) *l=3.873*

Fig. 2.: Effect of length scale on defusing the equivalent plastic strain concentration within shear band for 800 elements (*l* values are in μm)

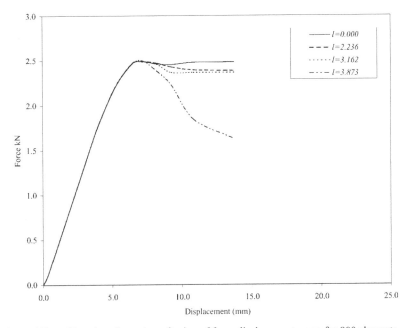

Fig. 3. Effect of length scale on the softening of force displacement curve for 800 elements (*l* values are in μm).

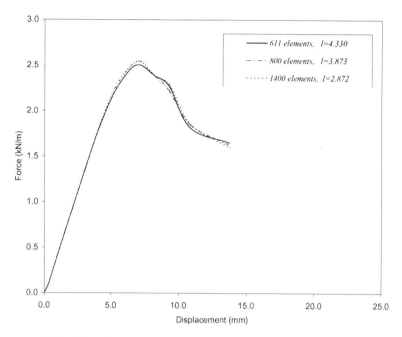

Fig. 4. Effect of length scale on the softening of force displacement curve for different mesh size and length scale values (l values are in μm).

A MICROMECHANICS MODEL FOR PLASTIC FLOW

The approach we adopt here is a generalization of a "double-shearing" plane strain constitutive model (Mehrabadi and Cowin 1978, Anand 1983, and Zbib 1991&93). The plane strain model is generalized to three–dimensions including the effects of elastic deformation, the typical pressure sensitive and dilatant hardening/softening response observed in granular materials, as well as the effect of heterogeneous microstructure. The model is motivated by crystal plasticity formulation. While in polycrystalline metals slip systems are uniquely defined, in soils active slip planes are determined by the state of stress and pre-stress conditions (anisotropy). In the double-slip model for soils, plastic deformation is viewed in terms of slip on two planes-defined by their normal unit vector n_i and slip direction m_i - and a phenomenological relation for the shear strain along each of the slip systems, such that:

$$D_{ij}^p = \sum_{s=1}^{2} \frac{\dot{\gamma}^{p(s)}}{2}(m_i^{(s)} \otimes n_i^{(s)} + n_i^{(s)} \otimes m_i^{(s)}) + \dot{v}^{(s)}(n_i^{(s)} \otimes n_i^{(s)}) \tag{26}$$

$$W_{ij}^p = \sum_{s=1}^{2} \frac{\dot{\gamma}^{p(s)}}{2}(m_i^{(s)} \otimes n_i^{(s)} - n_i^{(s)} \otimes m_i^{(s)}) \tag{27}$$

where D_{ij}^p is the plastic strain rate tensor, W_{ij}^p is the plastic spin, $\dot{\gamma}^{p(s)}$ is the shear strain rate on the *sth* slip system, and $\dot{v}^{(s)} = \beta\dot{\gamma}^{p(s)}$ is the dilatation rate, with $\beta(\gamma^p)$ being the *mobilized* dilatancy coefficient and γ^p is the effective plastic shear strain. Moreover, it is assumed that the slip planes are oriented at a given angle relative to the plane of maximum shear stress. This angle is then treated as a material parameter.

Further, the components of the elasticity tensor are assumed to be functions of the porosity density $\xi = $ *(void volume/total Volume)* whose evolution is given by:

$$\dot{\xi} = (1 - \xi) tr D^p \tag{28}$$

The hardening/softening behavior due to redistribution of contact is modeled through an evolution law for the *mobilized* friction coefficient; $\mu = \mu(\gamma^p, \xi)$ (e.g. Zbib and Aifantis 1989; Vardoulakis (1996), and Anand and Gu 2000). In passing we remark that Zbib (1993) showed that the above model is equivalent to a vertex-type plasticity model.

Since the above formulation does not have a length scale it cannot predict localization and patterning discussed earlier. Therefore, the theory presented above is modified to included gradient effects arising from the heterogeneous microstructure as proposed below.

Gradient theory – length scales

It is apparent, for the physical problem under consideration, that the main physical length-scale is the particle size. Furthermore, one can argue that there are two main physical mechanisms that seem to control flow, hardening/softening and localization, namely; i) particle-particle contact (magnitude and heterogeneous directional distribution) and, ii) porosity (magnitude and heterogeneous spatial distribution).

Modeling of heterogeneous directional-distribution

The magnitude of particle-particle contact is determined by the particle shape and relative orientation. During deformation redistribution of orientation takes place, which, in turn, causes changes in the magnitude of the mobilized friction coefficient. It can be argued that the change in the magnitude of contact area between two neighboring particles results from both shearing and relative local re-orientation. Within a continuum mechanics framework, local re-orientation between say two particles is attributed to changes in the local curvature. However, only the plastic component of the deformation tensor contributes to re-orientation (the elastic one is reversible). This distinction is very important as we discuss later. Thus, we define the rate of plastic curvature as:

$$\dot{\eta}_{ij} = \varepsilon_{jkl} (L^p)_{il,k}, \quad L_{ij}^p = D_{ij}^p + W_{ij}^p, L_{ij}^e = D_{ij}^e + \omega_{ij} \tag{29}$$

where ε_{jkl} is the permutation tensor. The quantity $\dot{\eta}_{ij}$ is a measure of the local plastic *"bending"* (This is analogous to the concept of dislocation density tensor in metal plasticity, Shizawa and Zbib 1999). Upon combining (29) with (26) & (27) we obtain:

$$\dot{\eta}_{ij} = \sum_{s=1}^{2} n_i \otimes (\nabla \dot{\gamma}^{p(s)})[n_i \otimes t_j - t_j \otimes n_i] + (\nabla \dot{v}^{(s)} \times n_i) \otimes n_j, \ t_j = n_i \times m_k \tag{30}$$

Modeling of heterogeneous porosity-distribution

The high-degree of heterogeneity of porosity within the representative volume element (*RVE*) cannot be ignored through simple homogenization assumptions. Suppose that the actual distribution of the porosity is known at every material point at the microscopic level (the particle length scale), say $\xi(x)$. Then the distribution of the porosity within the *RVE* is not uniform, since the scale of the *RVE* is a few orders of magnitude higher than that particle scale. The *RVE* scale can be defined as the *mesoscale*. Then, the average porosity over a meso-domain located at position x is given by:

$$\bar{\xi}(x) = <\xi> = \frac{1}{V}\int_V g(x-r)\xi(r)dV \quad V\in R-r,$$ (31)

where $(R-r)$ defines the size of the meso-domain and $g(x-r)$ is the distribution function. By expanding $\xi(r)$ around x we obtain:

$$\xi(r) = \xi(x) + \nabla\xi.(r-x) + \frac{1}{2!}\nabla^{(2)}\xi : (r-x)\otimes(r-x) +$$

$$\frac{1}{3!}\nabla^{(3)}\xi : (r-x)\otimes(r-x)\otimes(r-x) + \frac{1}{4!}\nabla^{(4)}\xi : (r-x)..$$ (32)

which when combined with (31) results in:

$$\bar{\xi}(x) = \xi(x) + c_1\nabla^2\xi + c_2\nabla^4\xi + ...$$ (33)

with c_1, c_2, being explicit functions of the size of the *RVE* (meso-domain). The theory is now completed by assuming that the *mobilized* friction coefficient is given by:

$$\mu = \mu(\gamma^p, \bar{\xi}, \bar{\eta})$$ (34)

where γ^p and $\bar{\eta} = \sqrt{\frac{1}{2}\eta_{ij}\eta_{ij}}$ are the effective plastic shear strain and effective plastic

curvature, respectively. It can be readily seen from (34) that there are two length scales included into the theory; one associated with the plastic curvature (orientation re-distribution) and the other related to the porosity re-distribution

The significance of using the plastic curvature as opposed to the total curvature goes beyond the fact that only the plastic component of deformation gives rise to permanent re-orientation. Following the theory developed by Shizawa and Zbib (1999), the model described can be developed within a thermodynamics framework and it can be shown that one obtains the classical set of balance laws with the stress tensor remaining symmetric.

Therefore, the model described above can be easily incorporated into a standard finite element program since the basic governing laws remain unchanged. The other main feature is that distinguishes the proposed constitutive model from past ones lies in the determination of microstructure parameters directly from experimental observations. These are currently under investigation.

CONCLUSIONS

This study demonstrates that incorporating a higher order gradient in terms of effective plastic shear strain into the flow stress of Mohr-Coulomb plasticity constitutive equation is effective in capturing the post localization behavior of granular materials. The gradient constitutive model has been incorporated into a finite element code. It was shown to simulate the biaxial shear test results on dry sand well. The predicted shear band thickness in a biaxial shear test was found to be in a good agreement with experimental results. The gradient term is found to act as a stabilizer and to eliminate the mesh size dependency on predicted stress strain behavior in the post localization regime where material exhibit strain softening.

Granular material consists of discrete particles and associated voids. Therefore, a higher order gradient constitutive model must explicitly account for its gradients in terms of the heterogeneous granular microstructure. Therefore, a higher order gradient continuum model that accounts for these mechanisms and associated length scales explicitly has been proposed. The model assumes that there are two main physical mechanisms that control flow, hardening/softening, and localization, namely, particle-particle contact (magnitude and heterogeneous directional distribution) and porosity (magnitude and heterogeneous spatial distribution).

ACKNOWLEDGMENT

The study presented in this paper was sponsored by the National Science Foundation under the grant CMS-0010124 to Washington State University. This support is gratefully acknowledged.

REFRENCES

ABAQUS, 2002. *Reference Manuals*. Hibbitt, Karlsson and Sorensen Inc, Pawtucket, RI.

Aifantis, E.C. 1987. The physics of plastic deformation. *Int.J.Plast*.3,211-247.

Anand, L. 1983, Plane Deformation of Ideal Granular Materials, *J. Mech. Phys. Solids*, 31 105-122.

Anand, L. and Gu, C . 2000. Granular Materials: Constitutive Equations and Strain Localization, *J. Mech. Phys. Solids*, 48, 8, p 1701-1733.

Balendran, B., Nemat_Nasser, S., 1993a. Double sliding model cyclic deformations of granular materials, including dilatancy effects. *J. Mech. Phys. Solids* 41 (3), 573-612.

Bazant, P. and Gambarova, B. 1984. Shear crack in concrete: Crack band microplane model, *J. Struct. Eng. ASCE*,110, 2015-2036 .

de Borst R. and Muhlhaus H. B. 1992. Gradient dependent plasticity: Formulation and algorithmic aspects, *Int. J. Numer.Meth. Engng.*, 35(3), 521-539.

Fleck NA, Hutchinson JW. 1993. A phenomenological theory for strain gradient effects in plasticity. *J. Mech. and Phys. Sol.*;41 :1825 –1857.

Han, C., Drescher, A., 1993. Shear bands in biaxial tests on dry coarse sand. Soils and Foundations 33, 118-132. *Int. J. Solids Structures*, 28, 845-857.

Mehrabadi, M. M. and Cowin, S.C. 1978. Initial Planar Deformation of Dilatant Granular Materials, *J. Mech. Phys. Solids*, 26, 269-284.

Mindlin R. D. 1964. Microstructure in linear elasticity, *Arch. Rational Mech. Anal.*, 10, 51-78.

Shizawa, K. and H.M. Zbib. 1999. Thermodynamical Theory of Strain Gradient Elastoplasticity with Dislocation Density: Part I - Fundamentals, *Int. J. Plasticity*, 15, 899-938.

Taylor, D. W. 1948. *Fundamentals of Soil Mechanics*. John Wiley . New York.

Vardoulakis, I. 1989. Shear banding and liquefaction in granular materials on the basis of a Cosserat continuum theory, *Ingenieur -Archiv*, 59, pp. 106-113.

Vardoulakis, I. 1996. Deformation of water-saturated sand: I. uniform undrained deformation and shear banding, *Geotechnique*, 46(3), pp. 441-456.

Zbib, H.M. 1991. On The Mechanics of Large Inelastic Deformations: Noncoaxiality, Axial Effects in Torsion and Localization, *Acta Mechanica*, 87, 179-196,

Zbib, H.M. 1993. On The Mechanics of Large Inelastic Deformations: Kinematics and constitutive modeling, *Acta Mechanica*, 96, 119-138.

Zbib, H.M. and Aifantis, E.C. 1988. On the Concept of Relative and Plastic Spins and Its Implications to Large Deformation Theories. Part II: Anisotropic Hardening Plasticity, *Acta Mechanica* 75, 35-56.

Zbib, H.M. and Aifantis, E.C. 1989. A Gradient-Dependent Flow Theory of Plasticity: Application to Metal and Soil Instabilities, *ASME J. Appl. Mech. Rev.*, 42, No. 11, Part 2, 295-304,.

Bifurcations & Instabilities in Geomechanics, – Labuz & Drescher (eds.)
© 2003 Swets & Zeitlinger, Lisse, ISBN 90 5809 563 0

FE-studies on formation of shear zones in granular bodies within a polar hypoplasticity

J. Tejchman
Gdańsk University of Technology, Poland

ABSTRACT: The paper deals with FE-investigations of shear localizations in granular bodies. Four different rate boundary value problems are numerically investigated: cyclic shearing of an infinite granular layer between two very rough boundaries, plane strain compression, passive and active movements of a retaining wall, and shearing in a direct and true simple shear apparatus. The FE-calculations are performed with a polar hypoplastic constitutive law using a mean grain diameter as a characteristic length.

1 INTRODUCTION

Failure in most engineering materials like metals, soils, polymers and concrete is proceeded by the occurrence of narrow zones of intense deformation. Due to these zones, a degradation of the material strength develops. The localizations of deformation can occur as cracks (if cohesive properties of the material dominate over frictional ones) or shear bands (if frictional properties are crucial). An understanding of the mechanism of the formation of these zone is of a crucial importance since they act as a precursor to ultimate fracture and failure. In order to properly analyze the failure behavior of materials, the phase of the localization of deformation has to be modeled in a physically consistent and mathematically correct manner. Classical FE-analyses within a continuum mechanics are not able to describe properly both the thickness of localizations zones and distance between them since they do not include a characteristic length of microstructure. Thus, they suffer from strong mesh sensitivity (its size and orientation) because differential equations of motion change their type and the rate boundary value problem becomes mathematically ill-posed. Thus, they require an extension by a characteristic length to accurately capture the formation of localizations. It can be achieved by the introduction in the continuum representation of strain gradients, polar terms, non-local quantities or viscosity.

The intention of the paper is to analyze shear localizations in granular bodies for the case of plane deformation. Four different rate quasi-static boundary value problems are dealt with: cyclic shearing of an infinite narrow layer between two very rough boundaries, shearing in a direct shear tester and true simple shear tester, plane strain compression test and earth pressures behind a retaining wall. The calculations were carried out with a polar hypoplastic model (Tejchman 1999, 2000, Tejchman & Gudehus 2001). This model was formulated within a Cosserat continuum by an extension of a non-polar model by polar quantities: rotations, curvatures, couple stresses and mean grain diameter. It can reproduce essential features of granular bodies during shear localization by taking into account the effect of the void ratio, pressure level, direction of deformation, mean grain diameter and grain roughness on the behavior of granular bodies. Due to the presence of a characteristic length, it can realistically capture the thickness and spacing of shear zones.

2 CYCLIC SHEARING

The calculations were performed with an infinite narrow granular layer during under conditions of free dilatancy (Tejchman & Bauer 2002). The calculations were carried out with only one element column with a height of h=20 mm and width of b=0.1 m, consisting of 20 quadrilateral horizontal elements. To eliminate the effect of sides, the so-called periodic boundary conditions were introduced, i.e. the displacements and Cosserat rotation along the sides were assumed to be free but constrained by the same value. As the initial stress state in the granular strip, a K_o-state without polar quantities was assumed. Gravity was neglected. The quasi-static deformation in sand was initiated through a constant horizontal displacement increments prescribed at the nodes along the top. Both bottom and top were very rough. The boundary conditions of the sand specimen were along the bottom: $u_1=0$, $u_2=0$, $\omega^c=0$, and along the top: $u_1=n\Delta u$, $\omega^c=0$, $\sigma_{22}=p$, where u_1 - horizontal displacement, u_2 – vertical displacement, ω^c – Cosserat rotation, n - number of the time step, Δu - displacement increment in one step, σ_{22} – vertical normal stress and p - vertical uniform pressure prescribed to the top. The specimen was first sheared in one direction up to $u_1^t/h =100\%$ to reach a residual (stationary) state (u_1^t – horizontal displacement of the top). Then the direction of shearing was repeatedly changed applying a large shear amplitude ($u_1^t/h=\pm200\%$).

Figures 1-5 show results for an initially dense sand strip (h=20 mm, e_o=0.60, d_{50}=0.5 mm, p =500 kPa, e_o – initial void ratio, d_{50} – mean grain diameter). Figures 1-2 present the evolution of normalized stress components σ_{ij}/h_s (h_s=190 MPa – granular hardness) at the mid-point of the strip (σ_{11} - horizontal normal stress in the direction of shearing, σ_{22} - vertical normal stress in the direction of shearing, σ_{33} – lateral normal stress perpendicular to the plane of deformation, σ_{12} - horizontal shear stress, σ_{21} - vertical shear stress), the evolution of the normalized vertical couple stress $m_2/(h_s d_{50})$ at the layer boundary and the evolution of the mobilized wall friction angle $\varphi_w=\arctan(\sigma_{12}/\sigma_{22})$ against the normalized horizontal displacement of the top u_1^t/h. The mobilized wall friction angle φ_w is related to the entire granular layer (stresses σ_{12} and σ_{22} are constant along both the height and length of the layer). Figures 3-4 demonstrate the evolution of the void ratio e in some points along the layer height versus u_1^t/h, and the distribution of the normalized couple stress $m_2/(h_s d_{50})$, the Cosserat rotation ω^c and the void ratio e along the normalized height x_2/d_{50} during the process of shearing. In addition, a deformed FE-mesh with a distribution of void ratio after initial shearing and after six full shear cycles is shown in Figure 5. A darker region indicates a higher void ratio.

All state variables (stress, couple stress and void ratio) tend to asymptotic values. The shear stresses insignificantly decrease during cyclic shearing. The shear stress σ_{12} is slightly larger than σ_{21} in the middle of the layer. During each reversal shearing, the normal stresses σ_{11} and σ_{33} diminish by 25%. The behavior of the couple stress m_2 is similar as for the shear stresses. The maximum wall friction angle is $42°$ (obtained during initial shearing). The residual wall friction angle is $30°$ and almost independent of the number of shear cycles. The void ratio close to the boundaries continuously decreases. The void ratio in the middle of the shear layer increases globally during each cycle and tends towards the pressure-dependent critical value ($e=e_c=0.75$). During each reversal shearing, contractancy occurs in the entire shear layer. After this additional compaction, the void ratio in the middle of the shear zone reaches again the critical value.

In the middle of the layer, the deformation is localized in a shear zone, characterized by the appearance of Cosserat rotations and a strong increase of the void ratio. At the boundaries of the shear zone, a strong jump of the curvature, stresses and couple stress takes place. The thickness of the shear zone, as visible from the Cosserat rotation and the couple stress jump at the shear zone edges, is about $14×d_{50}$ after the initial shearing and $18×d_{50}$ after six full shear cycles. Thus, during the cyclic shearing, the thickness of the shear zone grows. The increase of the thickness is pronounced within the first three shear cycles. The distribution of stresses σ_{11}, σ_{33} and σ_{21} across the shear zone is strongly non-linear. In the middle of the shear zone, the stresses σ_{11} and σ_{33} show their minimum and the stress σ_{21} its maximum. The stress ratios σ_{11}/σ_{22} and σ_{11}/σ_{33} become equal to 1.

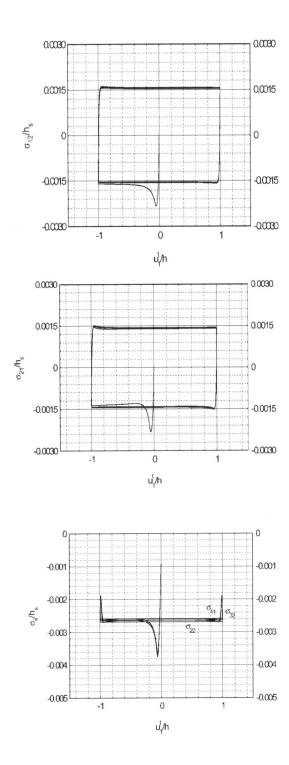

Figure 1. Evolution of normalized stresses σ_{ij}/h_s at the mid-point versus u_1^t/h

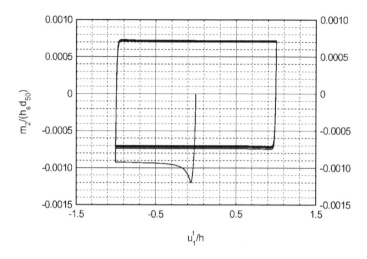

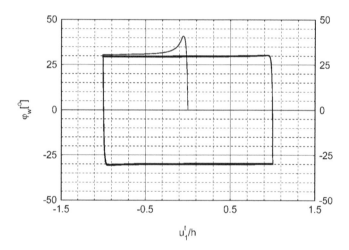

Figure 2. Evolution of normalized wall couple stress $m_2/(h_s d_{50})$ and mobilized wall friction angle φ_w versus u_1'/h

3 PLANE STRAIN COMPRESSION

The FE-calculations of a plane strain compression test were performed for a rectangular sand specimen with a height of h=50 mm and a length of l=100 mm (Tejchman 2002a). As the initial stress state in the granular specimen, a K_o-state was assumed. Two different sets of conditions along boundaries of the sand specimen were assumed: one with three non-deforming rigid boundaries and one free boundary, and second with four non-deforming rigid boundaries. In the first case, the granular specimen was placed on the smooth fixed bottom, its smooth top was subject to the uniform vertical pressure p, and its vertical smooth sides were subject to equal horizontal displacement increments directed to the specimen inside. In the second case, the granular specimen was also placed on the smooth fixed bottom. The vertical smooth sides were subject to equal horizontal displacement increments directed to the specimen inside, and the smooth top was subject

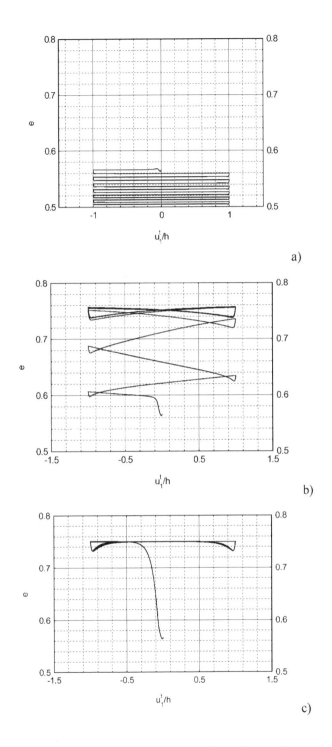

Figure 3. Evolution of void ratio e across the layer height versus u_1^t/h at x_2/d_{50}: a) 1, b) 13, and c) 19

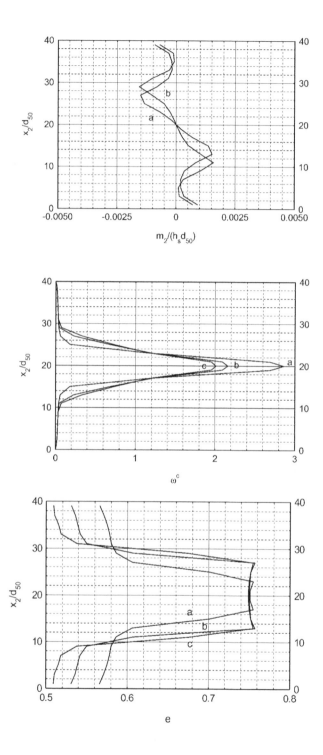

Figure 4. Distribution of normalized wall couple stress $m_2/(h_s d_{50})$ (a - after the initial shearing, b - after the sixth shear cycle), Cosserat rotation ω^c and void ratio e across the normalized height x_2/d_{50} (a - after the initial shearing, b - after the third shear cycle, c - after the sixth shear cycle)

a)

b)

Figure 5. Deformed FE-mesh with distribution of void ratio *e* for dense sand: a) after initial shearing, b) after the sixth shear cycle

to uniform vertical displacement increments directed to the specimen outside. The vertical displacement increments were equal to the horizontal ones. To trigger shear localization, the initial void ratio was distributed stochastically in the specimen elements by means of a random generator in such a way that the initial void ratio e_0 was increased in every element by the value $a \times r$, where a=0.01 or 0.05, and r is a random number within the range of (0.01, 0.99).

The effect of the stochastic distribution of the initial void ratio in the specimen is shown in Figure 6 (specimen with 3 rigid boundaries). The magnitude of the Cosserat rotation is marked by circles with a maximum diameter corresponding to the maximum rotation in the given step.

The results show that the geometry of shear localization is strongly influenced by the location of the first shear zone implied by the distribution of imperfections, and the magnitude of p. For the small vertical pressure of p=10 kPa, the number of shear zones is equal to four since the first two shear zones occur at the mid-point of the left side and one of them has a possibility to be reflected from a bottom. The thickness of shear zones is about $t=10 \times d_{50}$. If the uniform vertical pressure becomes greater (p=100 kPa), only one shear zone is created in the middle of the specimen since it has not a possibility to be reflected from rigid boundaries. The shape of this zone is slightly curved. However, if the prescribed vertical pressure increases linearly from 0 kPa to 100 kPa, three shear zones appear with a thickness of $t=16 \times d_{50}$. The distance between the shear zones is influenced by reflection positions of shear zones from the fixed bottom and rigid sides implied by the position of the first created shear zone.

The FE-calculations of a plane strain compression test with boundary conditions in the form of four non-deforming rigid boundaries and the stochastic distribution of the initial void ratio are shown in Figure 7. A pattern of shear zones is obtained since all boundaries can now reflect them. The geometry of shear zones is strongly influenced by factors e_0 and a. For a large deviation of the initial void ratio (a=0.05), different patterns with many intersecting shear zones are obtained with dense and medium dense sand.

a)

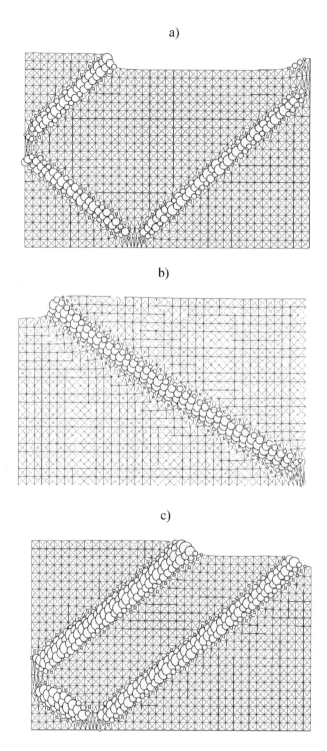

b)

c)

Figure 6. Deformed mesh with the distribution of the Cosserat rotation at residual state (u/l=0.05) for dense specimen (e_o=0.60+0.05r, d_{50}=0.5 mm): a) p=10 kPa, b) p=100 kPa, c) p=0-100 kPa

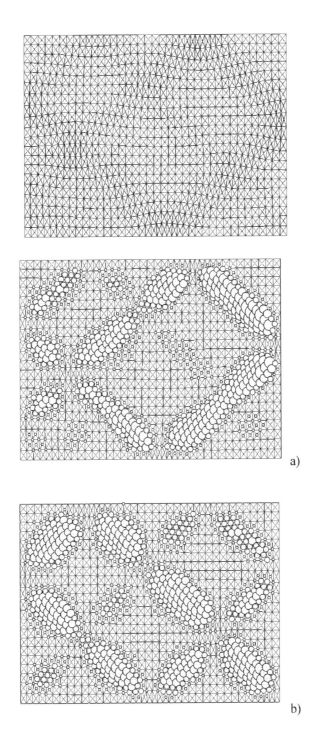

a)

b)

Figure 7.   Deformed mesh with the distribution of void ratio and Cosserat rotation at residual state (u/l=0.10) for dense specimen (d_{50}=0.5 mm): a) e_o=0.60+0.05r, b) e_o=0.65+0.05r

141

4 SHEARING WITH A DIRECT AND TRUE SIMPLE SHEAR APPARATUS

For simulations of two different shear box tests, a sand specimen with a length $l=0.10$ m and a height of $h=20$ mm was discretized with 2400 quadrilateral elements (Tejchman 2002b). As the initial stress state in the granular specimen, a K_o-state without polar quantities was assumed. A constant vertical load $p=100$ kN/m was prescribed along the entire top boundary. The bottom of the sand specimen was very rough ($u_1=0$, $u_2=0$ and $\omega^c=0$). Two side boundaries were assumed to be very smooth. To simulate a direct shear test, the same horizontal displacement increments were prescribed to both upper side boundaries: $u_1=n\Delta u_1$, $\sigma_{21}=0$, $m_1=0$ (Δu_1 – horizontal displacement increment, n – step number). In the case of a simple shear test, the conditions along side walls were: $u_1=n\Delta u_1(1-x_2/h)$, $\sigma_{21}=0$, $m_1=0$ (x_2 – vertical coordinate measured from the top). The horizontal displacements of all nodes along the top boundary were tied together ($u_1=n\Delta u_1$, $m_2=0$). In turn, the vertical displacements along the top boundary were constrained to move by the same amount.

In Figure 8, the deformed meshes with the distribution of the Cosserat rotation ω^c and void ratio e in the residual state are presented for direct and simple shearing. The distribution of the void ratio e is strongly non-uniform in the entire specimen during direct shearing. The thickness of a shear zone varies along the specimen lengths due to the effect of sides. During direct shearing, the Cosserat rotation is the largest at ends, during simple shearing, it is the largest in the middle of the specimen. The shear zone in a direct shear box test is slightly curved. The distribution of stresses is also strongly non-uniform in the entire specimen during direct shearing (Tejchman 2002b). In the case of simple shearing, the distribution is more uniform. The distributions of the Cosserat rotation ω^c and void ratio e at residual state along the vertical section in the center of the specimen are shown in Figure 9. In the middle of the sand specimen, a pronounced shear zone is created characterized by the occurrence of Cosserat rotations and an increase of void ratio. In the center of the specimen, it is approximately $t=6$-7 mm (12-$14\times d_{50}$) during direct shearing, and $t=6$-8 mm (12-$16\times d_{50}$) during simple shearing on the basis of the Cosserat rotation.

The evolution of the internal friction angle $\phi=\arctan(\sigma_{12}/\sigma_{22})$ versus the displacement u (at the mid-point of the specimen) are presented in Figure 10. The displacement u denotes the horizontal displacement of the upper half of the specimen during direct shearing or of the top boundary during simple shearing, σ_{12} denotes the horizontal shear stress, and σ_{22} is the vertical normal stress. The internal friction angle reaches a maximum for $u=0.5$ mm, shows a pronounced softening and tends to an asymptotic value for the horizontal displacement of the top boundary $u=8$-10 mm. The internal friction angles are higher for direct shearing than for simple shearing. During direct shearing, the internal friction angle, $\phi=\arctan(\sigma_{12}/\sigma_{22})$, is equal to $44°$ (at peak) and $35°$ (at residual state). In turn, the internal friction angle during simple shearing is equal to $42°$ (at peak) and $30.5°$ (at residual state). The shape of both curves is slightly different. The vertical displacement of the entire top boundary at $u=6$ mm is equal to 0.8 mm during direct shearing and 0.55 mm during simple shearing.

5 MOVEMENTS OF A RETAINING WALL

The calculations were performed for a sand body with a height of $H=200$ mm and a length of $L=100$ mm (Tejchman 2002c, d). The height of the wall was 42.5 mm and 170 mm. Totally, 3200 triangular elements were used. As the initial stress state in the granular specimen, a K_o-state without polar quantities was assumed ($\sigma_{22}=\gamma x_2$, $\sigma_{11}=\sigma_{33}=K_o\gamma x_2$, $\sigma_{12}=\sigma_{21}=m_1=m_2=0$); γ - initial density of sand, x_2 – vertical coordinate measured from the top, $K_o=0.40$ – pressure coefficient at rest. Two sides and the bottom of the sand specimen were assumed to be very rough ($u_1=0$, $u_2=0$, $\omega^c=0$). The top of the sand specimen was traction and moment free. The retaining wall was assumed to be stiff and very rough ($u_2=0$, $\omega^c=0$). Thus, no slip of sand along the wall was taken into account. The initial void ratio was distributed stochastically with a random generator.

In Figure 11, the deformed meshes with the distribution of the void ratio and Cosserat rotation in a residual state are shown for a passive mode (wall moves against the backfill). In the case of the wall translating horizontally (Fig.11a), four pronounced shear zones appear: one horizontal zone appearing at the wall bottom, one inclined zone spreading between the wall bottom and the free boundary and two radial shear zones appearing at the wall and connected to

A)

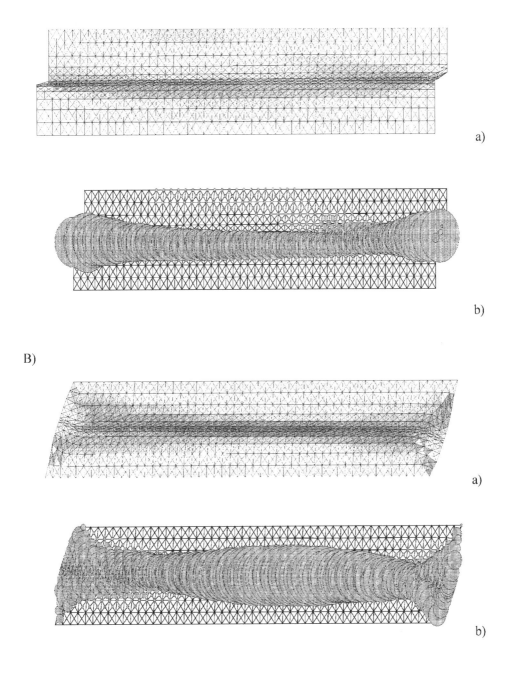

a)

b)

B)

a)

b)

Figure 8. Deformed FE-mesh with the distribution of void ratio e (a) and Cosserat rotation ω^c (b) during: A) direct shear test at the horizontal displacement of the upper specimen part u=6.0 mm, B) true simple shear test at the horizontal displacement of the top boundary u=10.0 mm

143

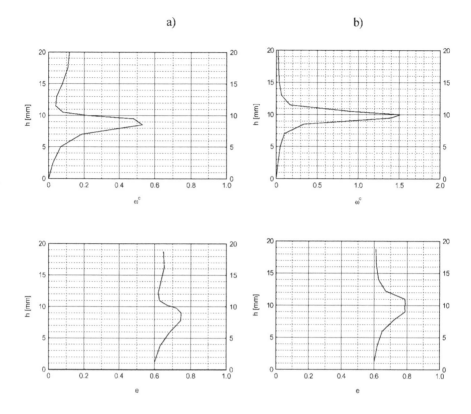

Figure 9. Distribution of Cosserat rotation ω^c and void ratio e in the vertical plane across the specimen center during: a) direct shear test at the horizontal displacement of the upper specimen part u=6.0 mm, b) true simple shear test at the horizontal displacement of the top u=10.0 mm

the inclined shear zone. The material starts to dilate at the same time at three different places (wall bottom, wall top and free boundary of the sand specimen). The second radial shear zone is not fully developed. The thickness of the inclined shear zone is about $20 \times d_{50}$ and its inclination from the bottom is about $\theta=40°$.

For a wall rotation around the bottom (Fig.11b), few parallel curved shear zones inside of the granular body are created. For a wall rotation around the top (Fig.11c), only one shear zone occurs which is more curved than the shear zone during the wall translation. The thickness of shear zones is similar.

The deformed meshes with the distribution of the void ratio and Cosserat rotation in a residual state for an active mode (wall moves away from the backfill) are presented in Figure 12.

In the case of the horizontal wall translation, only one interior shear zone is obtained. It is slightly curved. For a wall rotation around the bottom, two parallel interior shear zones occur. For a wall rotation around the top, one interior shear zone is again obtained. Its shape is parabolic.

The geometry of shear zones of Figures 11-12 is similar as in the experiments carried out at the Cambridge University and Karlsruhe University (Tejchman 2002d).

144

a)

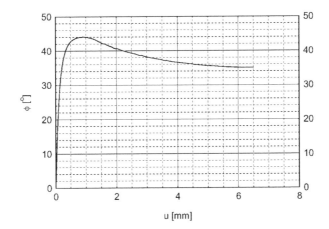

b)

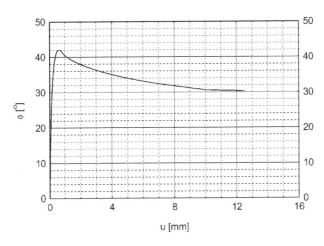

Figure 10. Evolution of the internal friction angle $\phi=\arctan(\sigma_{12}/\sigma_{22})$ in the middle of the sand specimen against the horizontal displacement of the top (a - direct shear test, b – true simple shear test)

6 CONCLUSIONS

The following conclusions can be drawn on the basis of FE-results within a polar hypoplasticity:

Cyclic shear amplitude influences the evolution of a localized shear zone. With an increase of the number of shear cycles, the thickness of a shear zone in an initially dense specimen increases.

If the shear layer is higher than the thickness of the localized zone, the void ratio beyond the localized zone decreases with each shear cycle.

Each change of the shear direction causes compaction in the shear layer.

The way of the shear inducement influences the strength of granular materials.

a)

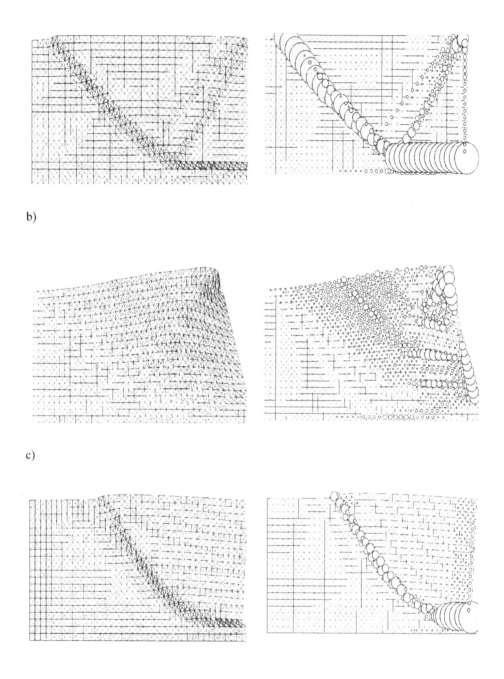

b)

c)

Figure 11. Deformed FE-meshes with distribution of void ratio and Cosserat rotation for dense sand (e_o=0.60) during passive earth pressure: a) translating wall, b) wall rotating around its bottom, c) wall rotating around its top

146

a)

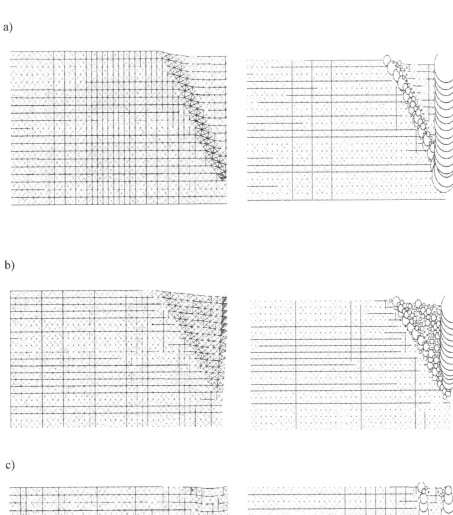

b)

c)

Figure 12. Deformed FE-meshes with distribution of void ratio and Cosserat rotation for dense sand ($e_o=0.60$) during active earth pressure: a) translating wall, b) wall rotating around its bottom, c) wall rotating around its top

147

The deformations and stresses are non-uniform during direct and simple shearing being affected by the type of shearing inducement, specimen size and boundary conditions along both sides. They are particularly non-uniform during a direct shear test.

The internal friction angles at peak and at residual state are higher during direct shearing than during simple shearing.

The thickness and shape of a shear zone appearing along the horizontal mid-section in a direct shear and true simple shear testers are non-uniform.

Shear zones have a tendency for reflection only from fixed or moving rigid boundaries.

The geometry of shear zones during a plane strain compression test depends mainly on the type of the boundary along the specimen (deforming or non-deforming boundary), specimen form and size, and the location of the first shear zone implied by the distribution and kind of imperfections. The distance between shear zones increases also with decreasing initial void ratio.

The thickness of shear zones increases with increasing initial void ratio, pressure level and mean grain diameter. If the initial void ratio approaches or exceeds the pressure-dependent critical void ratio, the shear zone reaches the size of the granular body.

The geometry of shear zones for the problem of earth pressures depends mainly on the type of the wall movement (active or passive, translation or rotation) and the specimen size.

The largest passive earth pressures occur with the horizontal translation of the wall, they are smaller with the wall rotation around the bottom and again smaller with the wall rotation around the top. The largest active earth pressures occur during the wall rotation around the top, the smallest ones during the horizontal translation.

Earth pressure coefficients from model tests cannot be directly transferred to large retaining walls due to a scale effect caused by the pressure level and mean grain diameter related to the specimen size.

The distribution of pressures along the wall is strongly non-linear in dense specimens.

REFERENCES

[1] Tejchman, J., Herle, I. & Wehr, J. 1999. FE-studies on the influence of initial void ratio, pressure level and mean grain diameter on shear localization. *Int. J. Num. Anal. Methods in Geomechanics*, 23(15): 2045-2074.

[2] Tejchman, J. 2000. Behaviour of granular bodies in induced shear zones. *Granular Matter*, Springer-Verlag, 2(2): 77-96.

[3] Tejchman, J. & Gudehus, G. 2001. Shearing of a narrow granular strip with polar quantities. *Int. J. Num. and Anal. Methods in Geomechanics*, 25: 1-28.

[4] Tejchman, J. & Bauer, E. 2002. Effect of cyclic shearing on shear localisation in granular bodies. *Granular Matter* (submitted for poublication).

[5] Tejchman, J. 2002a. Patterns of shear zones in granular materials within a polar hypoplastic continuum. *Acta Mechanica*, 155(1-2):71-95.

[6] Tejchman, J. 2002b. FE-simulations of a direct and a true simple shear test. *Powder Handling and Processing*, 14(2): 86-91.

[7] Tejchman, J. 2002c. Numerical investigations of shear localisations in earth pressure problems, *Proc. World Congress of Computational Mechanics, WCCM V*, July 7-12, 2002, Vienna, Austria. In H. A. Mang, F. G. Rammerstorfer & J. Eberhardsteiner (eds), http://wccm.tuwien.ac.at.

[8] Tejchman, J. 2002d. Evolution of shear localisation in earth pressure problems of a retaining wall. *TASK Quarterly*, Gdańsk University of Technology, 6(3): 387-411.

Bifurcations & Instabilities in Geomechanics, – Labuz & Drescher (eds.)
© 2003 Swets & Zeitlinger, Lisse, ISBN 90 5809 563 0

Constitutive modelling of granular materials with focus to microstructure and its effect on strain-localization

R.G. Wan
University of Calgary, Calgary, Alberta, Canada

P.J. Guo
McMaster University, Hamilton, Ontario, Canada

ABSTRACT: The paper presents a micromechanically-based constitutive model for frictional granular materials through a fabric (anisotropy) sensitive stress-dilatancy law. A second order tensor is used to describe fabric whose evolution during plastic deformation is chosen to be purely deviatoric and stress ratio dependent. This produces different fabrics at critical state that are consistent with the corresponding critical void ratio. To illustrate the role of fabric and its evolution, model simulations are presented for mixed drained-undrained stress paths within the context of instability. Shear band localization is also discussed with fabric as a controlling dependent variable. Model results suggest that, in an extreme case, a very loose sand could very well display post peak behaviour and shear banding due to fabric.

1 INTRODUCTION

The mechanical behaviour of a frictional granular material is strongly influenced by both fabric (anisotropy of grain packing) and grain contact force networks that develop during plastic deformations. When a sand body is sheared, an increase in volume (dilation) ensues as a consequence of geometrical constraints imposed by the fabric against applied stresses. This important phenomenon coined as stress-dilatancy hinges on particle kinematics (slip and spin) as the grains override each other against confinement. As such, both dilatancy and fabric control the nature of the deformation mode, e.g. the localization of deformations into a shear band, which usually signals incipient failure. It has been found that the dominant mechanism inside a shear band is that of particle rearrangement, including both rolling and translation leading into further fabric changes with respect to the region outside the shear band. In fact, Oda et al. (1998) and Desrues et al. (1996) both observed significant particle rotation and increase of voids within the shear band. Therefore, the proper description of stress-dilatancy with the inclusion of fabric information is a basic requisite for accurately modelling the stress-strain behaviour of sand leading to strain localization, see Wan & Guo (2001a).

Stress dilatancy theories based on macroscopic observations can be traced as far back to the early works of Rowe (1962). While the importance of confinement, density and stress path has been clearly demonstrated (Nakai, 1997; Houlsby 1991; Vaid and Sivathayalan, 2000; Pradhan et al. 1989; Gudehus 1996; Bauer and Wu 1993, and Wan & Guo 1999), experimental studies of stress dilatancy with focus to microstructural issues are scarce. In the realm of constitutive modelling, stress dilatancy equations used in conjunction with an elasto-plastic model abound in the literature, e.g. Nova & Wood (1979), Matsuoka (1974), Wan & Guo (1998). However, they are not powerful enough to describe aspects of sand behaviour related to its microstructure. Actually, most existing stress-dilatancy theories, except the notable works of Oda et al. (1998) among others, do not address microstructural issues probably due to the paucity of experimental studies in the literature. Recently, Wan & Guo (2001b,c) advocated a micromechanical approach in which continuum variables such as stress and strain are related to their microscopic counterparts, i.e. particle contact force and relative displacement respectively. By enforcing energy conservation at both micro- and macro-scales, a stress-dilatancy model describing the collective deformational behaviour of a system of rigid granules undergoing slips and rotations was formulated. It is also worthy to note the work of Nemat-Nasser (2000) which integrates micromechanics in a double sliding model.

In this paper, we use the model developed by Wan & Guo (2001b,c), which integrates a micromechanically based dilatancy rule, in order to illustrate the central role of fabric on granular material deformation with focus to both stress and strain paths as well as strain localization. The propensity of the fabric dependent constitutive model to succumb to shear band localization is examined by means of a bifurcation analysis. Hence shear band orientations and shear strains at localization can be computed under drained biaxial stress conditions for sand specimens with different initial fabrics. Model results demonstrate that very loose sand can display post peak behaviour and shear banding due to fabric.

2 MICROSTRUCTURALLY-BASED CONSTITUTIVE MODEL

We discuss only the essential features of the model in order to establish the background for the understanding of the constitutive modelling framework. For detailed treatise of mathematical developments, the reader is directed to Wan & Guo (1998, 1999, 2001b,c) and Guo (2000).

2.1 Stress-dilatancy with embedded fabric

In view of incorporating microstructural aspects into the formulation of stress-dilatancy, a representative elementary volume (REV) is chosen in which micro-variables are averaged and expressed in terms of macro-variables. For example, as a result of volume averaging, contact forces between particles can be expressed in terms of Cauchy stress, σ, via a so-called fabric tensor, F, that describes the geometrical arrangement of particles. Similarly, global strains ε can be linked to fabric and kinematical variables such as particle translations and rotations. Details can be found in Guo (2000) and Chang & Ma (1991).

The essence of the dilatancy model is to write energy dissipation considerations at grain contacts that slip and rotate during macroscopic deformations. For energy conservation at both scales, the rate of energy (power) dissipated $\dot{D}$ at the microscopic level must be equal to the work rate $\dot{W}$ expressed in terms of macro-variables, i.e.

$$\dot{W} = \sigma.\dot{\varepsilon} = \frac{1}{V} \int_{sliding} \dot{D}(\varepsilon) \, dV \tag{1}$$

over sliding contacts.

The power dissipated $\dot{D}$ can be further expressed in terms of micro-variables such as average tangential contact forces and relative slip displacements. As a result of such principle, a stress dilatancy equation with embedded micro-variables emerges in the form of a dilatancy rate β defined as the ratio of volumetric strain rate, $\dot{\varepsilon}_v$, over shear strain rate, $\dot{\gamma}$. These strain rates are wholly plastic. In conventional triaxial (axial symmetry) stress states, we get

$$\beta = -\frac{\dot{\varepsilon}_v}{\dot{\gamma}} = \frac{4}{3} \frac{(\sin \varphi_m - \sin \varphi_f)}{(1 - \sin \varphi_f \sin \varphi_m)} \tag{2}$$

with

$$\sin \varphi_f = \frac{X(F_{33}/F_{11}) + \gamma^*}{a^* + \gamma^*} \left(\frac{e}{e_{cr}}\right)^\alpha \sin \varphi_{cv} \tag{3}$$

in which e and e_{cr} are current and critical void ratios respectively, while φ_{cv} is the friction angle at critical state. Other parameters X, a^* and α are simply material constants. In Equation 2, the dilatancy rate depends principally on the relative difference between the mobilized friction angle φ_m and the characteristic friction angle φ_f that sets the threshold for zero dilatancy rate as well as the transition from contractancy to dilatancy. This threshold is not constant, but depends on a number of state parameters, principally fabric, as implied in Equation 3. More precisely, fabric information is embedded through fabric tensor components F_{11} and F_{33}, as well as a transformed plastic shear strain term γ^*, which is conventional strain factored with fabric as demonstrated in Guo (2000). The terms F_{11} and F_{33}, refer to the components of the fabric tensor associated with principal stress directions σ_1 and σ_3 respectively. In general the orientation of maximum density of contact normals coincides with the principal direction of the fabric tensor $\mathbf{F}$ and makes an angle θ with the major principal stress direction, as shown in Figure 1.

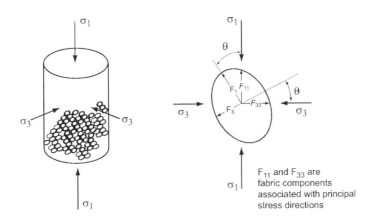

(a) sand specimen in axial symmetry subjected to principal stresses

(b) orientation θ of the major principal axis of fabric tensor F with principal stress direction

Figure 1. Sand specimen and fabric representation

It transpires from Equations 1 & 2 that both positive and negative rates of dilatancy can be obtained depending on the relative magnitudes of φ_m and φ_f. Thus, fabric conditions can be such that positive rate of dilatancy is possible even though the current void ratio is looser of critical, i.e. $e > e_{cr}$. As such, deformation rates engendered by such a stress-dilatancy equation reflect fabric dependencies when used as a flow rule in classical elasto-plasticity framework.

2.2 Elasto-plastic constitutive law

The constitutive model is formulated within a double yield surface elasto-plastic model describing both shear (s) and compaction (c) mechanisms. For describing the shearing mechanism, a linear Mohr-Coulomb yield condition f_s together with a plastic potential g_s are introduced, i.e.

$$f_s = (\sigma_1 - \sigma_3) - (\sigma_1 + \sigma_3)\sin\varphi_m = 0 \qquad (4)$$

$$g_s = (\sigma_1 - \sigma_3) - (\sigma_1 + \sigma_3)\sin\psi_m = 0 \qquad (5)$$

where a mobilized dilatancy angle ψ_m is introduced, with $\sin\psi_m$ being equal to the instantaneous dilatancy rate β defined in Equation 2. The linkage of Equation 5 with Equation 2 is essentially established through a flow rule in which the resulting deformation rate, $\dot{\varepsilon}^p = \dot{\Lambda}\,\partial g_s/\partial\sigma$, reflects fabric dependencies.

The evolution of yielding proceeds with the mobilization of friction angle that follows a coupled hardening/softening law, i.e.

$$\sin\varphi_m = \frac{\gamma^*}{a+\gamma^*}f_d(e)\sin\varphi_{cv}, \quad f_d(e)=\left(e/e_{cr}\right)^{-\beta} \qquad (6)$$

where α and β are material properties. Herein, the mobilization of friction angle, φ_m, is influenced by fabric by virtue of transformed plastic shear strain, γ^*, in Equation 6. The latter repre-

151

sents a hyperbolic variation of φ_m with γ^*, as well as total volumetric strains (deviatoric and isotropic pressure induced) through the void ratio function $f_d(e)$. It is also evident that, depending on whether the current void ratio e is denser or looser of critical, the factor $f_d(e)$ will adjust the functional representation of $\sin \varphi_m$ so as to make it evolve into either a softening or hardening trend with respect to γ^*.

Under isotropic stress conditions, the resulting compactive deformations are computed from a cap yield surface, f_c, which hardens isotropically in the stress space with the accumulation of irrecoverable volumetric plastic strains. The evolution of void ratio (resulting from both elastic and plastic deformations) with isotropic stresses is classically governed by an exponential law, i.e.

$$e = e_0 \exp\left[-\left(p^* / h_l\right)^m\right]$$
(7)

where h_l and m represent a modulus and an exponent number respectively, and transformed mean stress $p^* = \sigma.\mathbf{F}$. In subtracting the elastic component of the volumetric strain from the total one as given in Equation 7, the plastic volumetric strains due to compaction is obtained. The elastic component is calculated from a pressure and void ratio dependent bulk modulus K derived from the shear modulus G whose expression is

$$G = G_0 \frac{(2.17-e)^2}{1+e} \sqrt{p / p_{ref}}; \quad p_{ref} = 1\,\mathrm{kPa}$$
(8)

where G_0 is a material parameter, and a constant Poisson ratio ν is assumed. For further details about the origin and significance of the above expression see Wan & Guo (1999).

At critical state, the void ratio reaches its critical value, e_{cr}, which is by no means a constant, but varies with stress level and fabric. In fact, the attainment of a critical void ratio in a random arrangement of grains largely depends on the applied confining pressure as well as the geometrical shape of the grains. Thus, at the macroscopic level, an appropriate functional description of the critical void ratio dependency on stress level can be chosen as

$$e_{cr} = e_{cr0} \exp\left[-\left(\frac{p^*}{h_{cr}}\right)^n\right]$$
(9)

where e_{cr0} is the critical void ratio at very small confining stress, while h_{cr} and n are material parameters.

2.3 Fabric evolution

Finally, to complete the model, some evolution law must be introduced so as to describe the change in fabric during deformation history. For sake of simplicity, we assume that the rate of change of fabric tensor components $\dot{\mathbf{F}}$ is proportional to deviatoric stress ratio change $\dot{\eta}$. Defining the deviatoric stress ratio η as the deviatoric stress s over the mean stress p, the rate of fabric change becomes

$$\dot{\mathbf{F}} = \chi\dot{\eta} = \chi\left(\frac{\dot{s}}{p} - \frac{\dot{p}}{p^2}s\right)$$
(10)

where χ is a proportionality constant. The evolution law described in Equation 10 is purely deviatoric and precludes any fabric changes that would result under purely isotropic stress changes, a case which is not treated in this paper. Furthermore, Equation 10 implies coaxiality of $\mathbf{F}$ and $\dot{\eta}$, which is consistent with experimental observations where the fabric tensor strives to realign itself with the stress tensor, see Oda (1993). Furthermore, when the void ratio e reaches e_{cr} at critical state, $\dot{\mathbf{F}}$ approaches zero in order to ensure the upkeep of a constant stress ratio. Thus, at critical state, all components of the fabric tensor eventually achieve constant values. There would also be an alternative approach of using a deviatoric strain based fabric evolution law, but some physical inconsistency would be encountered, as fabric would not achieve a stationary value at critical state where strains are very large.

2.4 Incremental stress-strain relationship

When adopting plasticity theory described in the previous sections, we arrive at the incremental constitutive relationship:

$$\begin{bmatrix} C_{11} & C_{12} \\ C_{21} & C_{22} \end{bmatrix} \begin{Bmatrix} \dot{q} \\ \dot{p} \end{Bmatrix} = \begin{Bmatrix} \dot{\gamma} \\ \dot{\varepsilon}_v \end{Bmatrix}; \qquad \begin{aligned} \dot{q} &= \dot{\sigma}_1 - \dot{\sigma}_3, \; \dot{p} = (\dot{\sigma}_1 + 2\dot{\sigma}_3)/3 \\ \dot{\gamma} &= \dot{\varepsilon}_1 - \dot{\varepsilon}_3, \; \dot{\varepsilon}_v = (\dot{\varepsilon}_1 + 2\dot{\varepsilon}_3) \end{aligned} \tag{11a}$$

$$C_{11} = \frac{1}{3G} + \frac{1}{H_s} \frac{\partial f_s}{\partial q} \frac{\partial g_s}{\partial q}; \quad C_{12} = \frac{1}{H_s} \left[\frac{\partial f_s}{\partial p} - \frac{\lambda}{p} \frac{\partial f_s}{\partial e} \right] \frac{\partial g_s}{\partial q} \tag{11b}$$

$$C_{21} = \frac{1}{H_s} \frac{\partial f_s}{\partial q} \frac{\partial g_s}{\partial p}; \quad C_{22} = \frac{\lambda}{p(1+e_0)} + \frac{1}{H_s} \left[\frac{\partial f_s}{\partial p} - \frac{\lambda}{p} \frac{\partial f_s}{\partial e} \right] \frac{\partial g_s}{\partial p} \tag{11c}$$

in which λ = material parameter, e_0 = initial void ratio and H_s = plastic hardening modulus.

3 SIMULATION OF MIXED DRAINED-UNDRAINED LOADING PATHS

It has been brought to attention (Laoufa & Darve, 2002 and Chu & Leong, 2001) the existence of loading paths in the triaxial stress condition ($\sigma_2 = \sigma_3$) whereby the failure stresses are well below the conventional failure surface of the Mohr-Coulomb type. For example, the effective stress path of a loose sand tested in undrained triaxial compression reaches a peak point (well below the failure surface) at, and after which, controllability is lost since the second order work $d^2W = d\sigma.d\varepsilon \leq 0$. The study of such constitutive feature is of interest from both the viewpoints of material instability and fabric changes (inherent and induced anisotropy). It is found that the material instability that occurs at the peak of the effective stress path does not correspond to shear band localization typical of a dense sand response in drained conditions. As such, Chu et al. (1992) refers to failure at the peak point as a `pre-failure' strain softening which is reckoned to be a true material property significantly affected by both fabric and effective confining pressure.

In this section, we investigate the effect of fabric changes for a range of stress paths in relation to the peak point alluded in the above, and which is unstable. Indeed, it is postulated that the fabric state at the peak point is highly anisotropic and defines a bifurcation point at which the material behaviour drastically changes from one form to another. This argument is illustrated by analyzing a series of mixed stress paths tests in which the material is pre-sheared along a drained path (CD test) up to a stress ratio $R = \sigma_1/\sigma_3$, and deformations are thereafter switched to undrained conditions (CU test).

3.1 Effective stress paths in mixed CD/CU tests

In the numerical examples given below, we simulate a mixed drained-undrained stress path on a sand specimen loose of critical with an initial fabric inherited from moist tamping during specimen preparation. The initial fabric corresponds to horizontal bedding planes ($\theta=0^0$) with a bias of contacts in the axial direction. A ratio of axial to radial fabric components is arbitrarily chosen as $F_1/F_3 = 1.1/0.95$ for illustration purposes. The sand is considered to be loose with an initial void ratio of $e_0 = 0.8$. Also, the purely deviatoric evolution law given by Equation 10 is assumed for fabric component changes ($\dot{F}_{ij}$) during deformation history.

3.2 Effect of pre-shear (stress-ratio)

First, we consider an undrained test (CU2) with an initial effective consolidation stress of 410 kPa, and compute the effective stress path response as represented by ABC, where the peak point B coincides with the instability point. The so-called instability line with a slope $\eta^* = q_B/p_B$ as defined by Lade (1993) can be identified, see Figure 2a. Next, a drained (CD) test is simulated using the same initial void ratio and fabric conditions such that the drained stress path passes through the peak point B on the effective stress path of the previous test. The idea is to pre-shear the material along the drained path at different deviatoric stress levels (corresponding to stress ratios $R = 1.2, 1.5, 1.8, 2.0, 2.35$ and 3.0), and thereafter switch to undrained deformations until failure. Figures 2a, b show the effective stress paths and associated stress-strain responses computed by the model.

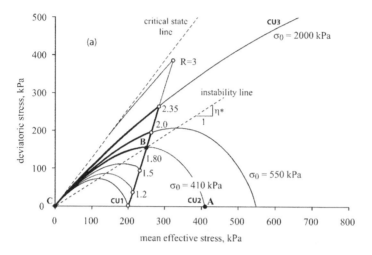

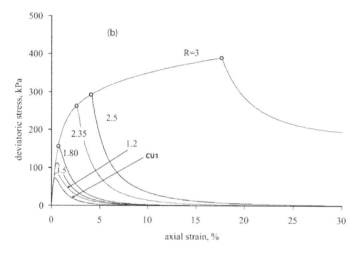

Figure 2. Simulation of mixed drained/undrained response of loose sand with horizontal bedding plane: (a) effective stress path, (b) stress-strain curves

It is observed that the undrained effective stress path AB, together with line segment BC with slope $\eta = \eta^*$, define an envelope inside which the material behaviour is stable. For example, if the drained path switches into an undrained one at a stress state within the envelope, the subsequent effective stress path displays an initial increase in deviatoric stress (strength), while a sharp drop in deviatoric stress is obtained if the stress state lies outside the envelope. In all situations, the deviatoric stress ratio η increases monotonically until it reaches its limit value η_{cv} at critical state. If the value of η at the drained-undrained transition is η^*, then as long as $\eta < \eta^*$, the undrained stress paths will display predominantly stable behaviour corresponding to work hardening. However, if $\eta > \eta^*$, i.e. point of drained-undrained transition lies above the instability point, the subsequent undrained path is very unstable and follows a descending branch. Thus, point *B* can be regarded as a bifurcation point associated to undrained conditions.

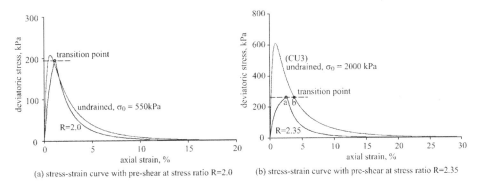

(a) stress-strain curve with pre-shear at stress ratio R=2.0

(b) stress-strain curve with pre-shear at stress ratio R=2.35

Figure 3. Effect of pre-shear on subsequent stress-strain responses

The undrained effective stress path in a pre-sheared test tends to follow quite closely that of the purely undrained test (see $R = 2.0$ and 2.35 compared with CU2 and CU3 on Figure 2a). However, when looking at deformations (Figures 3a, b), pre-shearing leads to a more brittle behaviour (sharp drop of deviatoric stress with axial strain) than in the purely undrained case. Also, the axial strain for the pre-sheared case at the transition point is different from the one in the purely undrained case. The differences in stress-strain behaviours depicted in Figures 3a, b are due to the fact that both the void ratios and fabrics are different for the same stress point. The difference in principal fabric ratios achieved at the same stress point ($R = 2.35$) in the two stress paths considered is clearly shown at points *a* and *b* in Figure 4 below.

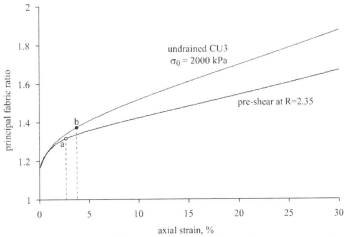

axial strain, %

Figure 4. Comparison of fabric evolutions for undrained and drained pre-shear/undrained paths

More precisely, the model calculations indicate that more fabric is being developed along the undrained stress path than during the pre-shearing phase in the drained stress path. This is in line with the physics of the problem given that much fabric changes through particle reorganization are expected to occur under isochoric conditions than in a drained test. Figure 5 shows the evolution of principal fabric ratio with axial strains for different amounts of pre-shearing. The fabric developed in the pre-shearing stage seems to be masked by the one occurring in the subsequent undrained path, especially when pre-shearing occurs at small stress ratios. The rate of fabric change with deformations is in fact controlled by a number of factors such as level of deviatoric stress ratio η, and rate of change of both mean effective and deviatoric stresses.

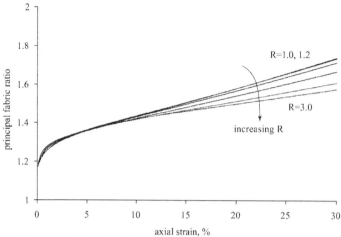

Figure 5. Fabric evolutions for different drained pre-shear levels

3.3 Effect of fabric evolution rate

The fabric evolution rate was next investigated in order to access its role when the stress path is switched from drained to undrained conditions at a given level of pre-shear, here a stress ratio $R = 1.8$. Figure 6a shows the effective stress paths obtained for various conditions, including the purely undrained one. Although Figure 6a shows that the slow fabric evolution rate ($\chi = 0.01$ in Equation 10) case gives an effective stress path close to the one obtained for the purely undrained case, the corresponding stress-strain responses are quite different as illustrated in Figure 6b. The fast fabric evolution rate corresponds to a ten-fold increase in χ, which produces a remarkable change in effective stress path as shown in Figure 6a. During the course of pre-shearing, fast fabric evolution results into stronger fabric in the vertical direction (i.e. the direction of σ_1) at a given deviatoric stress, leading to larger deformation stiffness. More rapid softening was also obtained during undrained shearing with fast fabric evolution. This numerical exercise illustrates the pivotal role of fabric and its evolution rate on material response.

4 SHEAR BAND ANALYSIS

A shear band analysis is carried out on the elasto-plastic constitutive model described in the previous section. Biaxial stresses are considered in the analysis since most experimental strain localization data are available for these conditions. Thus, the stress state σ of a granular material under biaxial stress conditions is typically given by

$$\sigma = \begin{bmatrix} \sigma_1 & 0 \\ 0 & \sigma_2 \end{bmatrix} ; \quad \sigma_1 > \sigma_2 > 0 \tag{12}$$

in which positive stresses refer to compression. The following definitions are recalled for biaxial stress conditions for later use in the shear band analysis. Let $p = \sigma_{kk}/2$, $\tau = (s_{ij}s_{ij}/2)^{1/2}$, $\varepsilon_v = \varepsilon_{kk}$, $\gamma^p = (2e_{ij}^p e_{ij}^p)^{1/2}$, $\mu = \sin \varphi_m = \tau/p$, $\beta = \sin \psi_m = -d\varepsilon_v^p/d\gamma^p$. Furthermore, s_{ij} and e_{ij}^p represent deviatoric stress and deviatoric plastic strain tensors respectively. Also repeated indices which vary from 1 to 2 refer to summation.

In order to carry out a stability analysis on the proposed constitutive model, the incremental stress-strain equations ($\sigma_{ij} = C_{ijkl}\varepsilon_{kl}$) must first be established. These are typically obtained by applying plastic consistency conditions together with the flow rule based on the stress-dilatancy equation described earlier. Typically, the fourth order constitutive tensor then emerges as

$$C_{ijkl} = G\left(L_{ijkl}^e + L_{ijkl}^p\right) \tag{13}$$

156

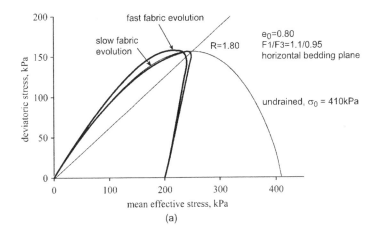

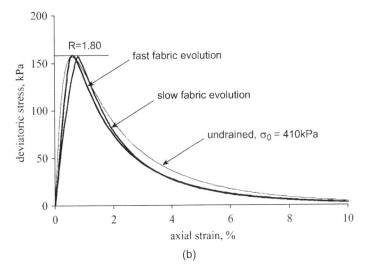

Figure 6. Effect of fabric evolution rate: (a) effective stress path, (b) stress-strain response

where

$$L_{ijkl}^c = (\Theta - 1)\delta_{ij}\delta_{kl} + 2\delta_{ik}\delta_{jl} \tag{14}$$

with $\Theta = (1 + \nu)/(1 - \nu)$ and $\nu =$ Poisson ratio.

The plastic part of the constitutive tensor is given as

$$L_{ijkl}^p = -\frac{<1>}{H}\left(\frac{s_{ij}}{\tau} + \Theta\beta\,\delta_{ij}\right)\left(\frac{s_{kl}}{\tau} + \Theta\delta_{kl}\right) \tag{15}$$

The operator $< \ >$ represents a switch function, and H is a plastic hardening modulus defined as

$$H = 1 + h + \Theta \beta \mu \quad \text{and} \quad h = \frac{p \; d\mu}{G \; d\gamma^p} \tag{16}$$

The condition for strain localization into a shear band is well established as it can be obtained from equilibrium of stress rates across a possible shear band. For example, in the case of continuous bifurcation and small material rotations, the condition is given as $\det(\mathbf{n}:\mathbf{C}:\mathbf{n}) = 0$ (Rice and Rudnicki, 1980). The orientation $\mathbf{n}$ of the normal to the shear band is calculated as real roots of $\det(\mathbf{n}:\mathbf{C}:\mathbf{n}) = 0$. In the case of biaxial stress, the localization condition for flow plasticity theory can be written as

$$(C - a_1 C_{2211} - a_2 C_{1122})^2 - 4 a_1 a_2 \, C_{1111} C_{2222} = 0 \tag{17a}$$

$$C = C_{1111} C_{2222} - C_{1122} C_{2211} \tag{17b}$$

$$a_1 = \frac{1}{2} C_{1212} + \sqrt{s_{ij} s_{ij} / 2} \; ; \; a_2 = \frac{1}{2} C_{1212} - \sqrt{s_{ij} s_{ij} / 2} \tag{17c}$$

The shear band inclination α with respect to the horizontal is finally obtained as the roots of

$$\tan \alpha = \pm \sqrt[4]{(a_2 C_{2222})/(a_1 C_{1111})} \tag{18}$$

5 INFLUENCE OF FABRIC ON SHEAR BAND LOCALIZATION IN PLANE STRAIN

We examine shear band localization using the bifurcation condition described in Equations 16 & 17 applied to the fabric based constitutive model presented in Section 2. In all following numerical simulations of plane strain compression, the specimen is initially in hydrostatic stress state. Shearing is simulated by imposing shear strains, $\gamma = \varepsilon_1 - \varepsilon_2$, to the specimen with horizontal stresses being kept constant, while both vertical and out-of-plane stress components increase depending on the pressure, current density and fabric according to the proposed constitutive model.

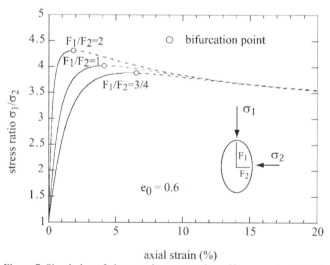

Figure 7. Simulation of plane-strain compression of loose sand with different initial fabrics

Figure 7 shows stress-strain as well as volumetric strain curves for plane strain compression simulations with points of bifurcation signalling possible shear band localization. A loose sand is considered with $e_0 = 0.6$ starting at the hydrostatic state with confining stress of

158

$\sigma_c = 100\text{kPa}$. The fabric tensor is defined by its two components F_1 and F_2, whereas its orientation is given by the angle θ between the major principal stress σ_1 and the major fabric component F_1. Simulations are carried out such that for a given initial void ratio and a fixed initial fabric orientation ($\theta = 0^0$), the ratio of major to minor fabric components ($\Omega = F_1 / F_2$) is varied so as to describe various degrees of initial anisotropy. The general trend of results in Figure 7 indicates that, as the degree of anisotropy is increased, shear band strain localization becomes more conducive as the shear strain at bifurcation gets smaller. Also, it is interesting to note that as the degree of initial anisotropy increases, post peak behaviour is more pronounced.

Next, we consider the localization behaviour of sand subjected to an initial confining stress of 100 kPa with varying initial principal fabric and void ratios. The initial orientation of the fabric tensor is however kept at zero degrees. Figure 8a indicates a general decrease in axial strain at localization with increasing initial principal fabric ratio (anisotropy) for a fixed initial void ratio. For initial void ratios greater than 0.7 (very loose states), localization is less likely to occur even by increasing initial anisotropy because predicted axial strains at localization are quite large. It is also found in Figure 8b that shear band inclinations become shallower for looser states, while larger anisotropies in the vertical direction lead to slightly steeper shear bands.

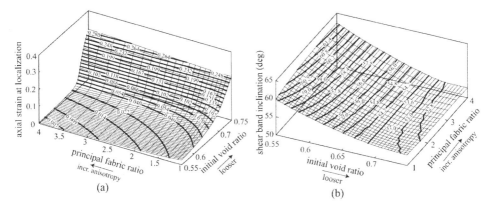

Figure 8. Effect of initial void ratio and principal fabric ratio on localization ($\theta = 0$ and $\sigma_0 = 100\text{kPa}$): (a) axial strain at localization and, (b) shear band inclination.

6 CONCLUSIONS

An elasto-plastic constitutive model incorporating microstructural information through a stress-dilatancy law was presented. A shear band analysis was also applied to the incremental constitutive equation in view of examining the effect of fabric on shear band localization. Stress paths in which the material was pre-sheared in drained conditions at a given stress ratio, and thereafter sheared under undrained conditions were simulated with the consideration of fabric and rate of fabric evolution. It was found that there exists a critical magnitude of pre-shear above which the undrained effective stress path becomes unstable. The amount of pre-shear in drained conditions also control the extent of induced fabric change that subsequently governs the stability of the subsequent undrained effective stress path. This mixed drained/undrained stress path find its application in soil-structure interaction problems under extreme loading conditions in which pre-shear stress always exists before a granular soil mass reaches its ultimate deformation state. Finally, the effect of fabric on strain localization propensity was investigated. It was found that, for a given void ratio and stress level, the higher the degree of anisotropy (principal fabric ratio), the easier it is to trigger strain localization and the smaller the shear strain at which this happens. In extreme cases, it could happen that a loose sand shows propensity to strain localization with post-peak behaviour. Finally, we think that the framework of dilatancy and fabric as presented in this paper deserves a thorough examination, if better insights into the stress-strain behaviour of granular soils are to be gained.

7 ACKNOWLEDGEMENTS

The authors wish to acknowledge funding provided by the Natural Science and Engineering Research Council of Canada.

REFERENCES

Bauer, E. & Wu, W. 1993. A hypoplastic model for granular soils under cyclic loading. *Proceedings of the International Workshop on Modelling Approaches to Plasticity*, Elsevier, 247-258.

Chang, C.S. & Ma, L. 1991. A micromechanical-based micropolar theory for deformation of granular solids. *Int. J. Solids Structures* 28(1), 67-86.

Chu, J. & Leong, W.K. 2001. Pre-failure strain softening and pre-failure instability of sand: A comparative study. *Géotechnique* 51(4): 311-321.

Chu, J., Lo, S-C.R., & Lee, I.K. 1992. Strain-softening behaviour of granular soil in strain-path testing. *J. Geotech. Eng., ASCE* 118(2): 191-208.

Desrues, J., Chambon, R., Mokni, M. & Mazerolle, F. 1996. Void ratio evolution inside shear bands in triaxial sand specimens studied by computed tomography. *Géotechnique* 46(3): 529-546.

Gudehus, G. 1996. A comprehensive constitutive equation for granular materials. *Soils and Foundations* 36(1), 1-12

Guo, P. J. 2000. Modelling granular materials with respect to stress-dilatancy and fabric: A fundamental approach. Ph.D. Dissertation, Civ. Engrg. Dept., University of Calgary, 375 p.

Houlsby, G.T. 1991. How the dilatancy of soils affects their behaviour. *Proc. 10th Eur. Conf. Soil Mech. Found. Eng.* 4: 1189-1202.

Lade, P.V. 1993. Initiation of static instability in the submarine Nerlerk berm. *Canadian Geotechnical Journal* 30: 895-904.

Laouafa, F. & Darve, F. 2002. Modelling of slope failure by a material instability mechanism. *Computers and Geotechnics* 29 (4): 301-325.

Matsuoka, H. 1974. Dilatancy characteristics of soils. *Soils and Foundations* 14(3): 45-53.

Nakai, T. 1997. Dilatancy characteristics of geomaterials. In: *Deformation and Progressive Failure in Geomaterials*, IS-Nagoya'97 (Asaoka, A, Adachi, T. and Oda, F. Eds), 899-906.

Nemat-Nasser, S. 2000. A micromechanically-based constitutive model for frictional deformation of granular materials. *Journal of the Mechanics and Physics of Solids* 48: 1541-1563.

Nova, R. & Wood, D.M. 1979. A constitutive model for sand in triaxial compression. *Int. J. Num. Analy. Meth. Geomech.* 3: 255-278.

Oda, M. (1993). Inherent and induced anisotropy in plasticity of granular materials. *Mechanics of Materials* 16: 35-45.

Oda, M., Kazama, H. & Konishi, J. 1998. Effects of induced anisotropy on the develpoment of shear bands in granular materials. *Mechanics of Materials* 28: 103-111.

Pradhan, T.B.S., Tatsuoka, F. & Sato, Y. 1989. Experimental stress dilatancy relations of sand subjected to cyclic loading. *Soils and Foundations* 29: 45-64.

Rice, J. & Rudnicky, J.W. 1980. A note on some features on the theory of localization of deformation. *Int. J. Solids Structures* 16, 597-605.

Rowe, P.W. 1962. The stress-dilatancy relation for static equilibrium of an assembly of particles in contact. *Proc. of the Royal Society of London*, Vol 269, Series A, 500-527.

Wan, R.G. & Guo, P.J. 1998. A simple constitutive model for granular soils: Modified stress-dilatancy approach. *Computers and Geotechnics* 22(2): 109-133.

Wan, R.G. & Guo, P.J. 1999. A pressure and density dependent dilatancy model for granular materials. *Soils and Foundations* 39(6): 1-12.

Vaid, Y.P. & Saivathayalan, S. 2000. Fundamental factors affecting liquefaction susceptibility of sands. *Can. Geotech. J.* 37: 592-606.

Wan R.G. & Guo P.J. 2001a. Effect of fabric on strain localization of sands. *Proceedings of the 1st. Asia-Pacific Conference on Computational Mechanics*, Nov. 20-23, 2001, Sydney, Australia, (1): 503-508.

Wan R.G. & Guo P.J. 2001b. Effect of microstructure on undrained behaviour of sands. *Can. Geotech. J.* 38: 16-28.

Wan R.G. & Guo P.J. 2001c. Drained cyclic behaviour of sand with fabric dependence. *Journal of Engineering Mechanics, ASCE* 127(11): 1106-1116.

160

Bifurcations & Instabilities in Geomechanics, – Labuz & Drescher (eds.)
© 2003 Swets & Zeitlinger, Lisse, ISBN 90 5809 563 0

Influence of initially transverse isotropy on shear banding in granular materials

E. Bauer
Institute of General Mechanics, Graz University of Technology, Austria

W. Wu
Electrowatt Infra AG, Zürich, Switzerland

W. Huang
Department of Civil, Surv. & Env. Eng., The University of Newcastle, Australia

ABSTRACT: This paper focuses on the analysis of shear band bifurcation in a cohesionless and initially transversely isotropic granular material based on a hypoplastic constitutive model. The constitutive equation for the evolution of the stress is formulated with a nonlinear tensor-valued function depending on the current void ratio, the Cauchy stress, the rate of deformation and a structure tensor for anisotropy effects. A decrease of the effect of anisotropy as a result of a reorientation of particles during dilatant deformation is assumed. It is shown that for homogeneous compression the incremental stiffness and the mobilized peak friction angle strongly depend on the initial void ratio and the orientation of the isotropic plane. The possibility of shear band formation is investigated and the influence of the state parameters on the shear band inclination is discussed.

1 INTRODUCTION

In natural sand deposits an initially transverse isotropy can be explained by a preferred orientation of the long axis of non-spherical particles as a result of the sedimentation process (e.g. Oda et al. 1985). The plane perpendicular to the deposit direction is called bedding plane and it is a plane of isotropy. Experimental studies with sand specimens show that the orientation of the bedding plane relative to the principal stress directions has a significant influence on the stress-strain behaviour (e.g. Arthur & Phillips 1975, Lam & Tatsuoka 1988, Tatsuoka et al. 1990). The stiffness and the peak friction angle are higher for loading perpendicular to the bedding plane than for loading parallel to it. They are also influenced by the mean pressure and the current density. However, for large monotonic shearing the stress-strain curve after the peak approaches a stationary state, where the stress ratio seems to be independent of the bedding plane (e.g. Yamada & Ishihara 1979). This indicates that under large shearing the initial anisotropy declines as a result of grain rotations and grain rearrangements and it may be swept out when the granular material reaches a critical state (e.g. see Figure 13 in Lam & Tatsuoka 1988).

The focus of the present paper is on modelling the mechanical behaviour of an initially transverse isotropy in dry and cohesionless granular materials based on a continuum approach. For this purpose a particular hypoplastic constitutive model by Gudehus (1996) and Bauer (1996) for a cohessionless and initially isotropic material is extended with respect to transverse isotropy. The present hypoplastic model takes into account the current void ratio, the Cauchy stress tensor, the rate of deformation, and a structure tensor defining the isotropic plane. In order to model the non-linear and inelastic material behaviour the evolution equation for the stress tensor consists of the sum of a tensor function which is linear in the rate of deformation and a tensor function which is non-linear in the rate of deformation according to the concept of hypoplasticity (Kolymbas 1988, 1991). Concerning a more general classification, the hypoplastic model can be assigned to a class of constitutive models referred as incrementally non-linear models (Darve 1991). Transversely isotropic material properties are included with a certain invariant form of a tensor function given by Boehler & Sawczuk (1977). The tensor function depends on the stress tensor and on the second order structure tensor and it is incorporated in the nonlinear part of the constitutive equation according to the concept proposed by Wu (1998). While the coefficients of anisotropy are assumed to be constant in the earlier version by Wu (1998), an evolution of anisotropy depending on the volume strain rate is considered in the present paper. It is assumed that the influence of the initial anisotropy decreases for large shearing and it will be swept out in stationary or so-called critical-states.

The paper is organized as follows: For an isotropic material behaviour the concept of hypoplasticity is briefly outlined in Section 2.1. The constitutive model given by Bauer (1996) and Gudehus (1996) is extended to transversely isotropic material according to the proposal by Wu (1998), Bauer & Huang (1999) in Section 2.2. In Section 3 the possibility of spontaneous shear band formation is studied based on the general bifurcation theory given by Hill (1962), Rudnicki & Rice (1975), Rice & Rudnicki (1980). The bifurcation conditions are derived in a way similar to the ones outlined for isotropic hypoplastic material models in earlier publications (e.g. Chambon & Desrues 1985, Wu & Sikora 1991, 1992, Bauer 1999, Tamagnini et al. 2000, Wu 2000). For the present model the bifurcation condition is investigated under plane strain conditions.

Throughout the paper compression stress and strain are defined as negative. Bold lower case, bold upper case and calligraphic letters denote vectors, tensors of second order and of fourth order, respectively. In particular, the identity tensor of second order is denoted by $\mathbf{I}$ and the identity tensor of fourth order is denoted by $\mathcal{I}$. For vector and tensor components indicial notation with respect to a rectangular Cartesian basis $\mathbf{e}_i$ $(i = 1, 2, 3)$ is used. Operations and symbols are defined as: $\mathbf{a}\,\mathbf{b} = a_i\,b_i$, $\mathbf{A}\,\mathbf{b} = A_{ij}\,b_j\,\mathbf{e}_i$, $\mathbf{a} \otimes \mathbf{b} = a_i\,b_j\,\mathbf{e}_i \otimes \mathbf{e}_j$, $\mathbf{A} \otimes \mathbf{B} = A_{ij}\,B_{kl}\,\mathbf{e}_i \otimes \mathbf{e}_j \otimes \mathbf{e}_k \otimes \mathbf{e}_l$, $\mathbf{A}\mathbf{B} = A_{ik}\,B_{kj}\,\mathbf{e}_i \otimes \mathbf{e}_j$, $\mathcal{A} : \mathbf{B} = A_{ijkl}\,B_{kl}\,\mathbf{e}_i \otimes \mathbf{e}_j$ and $\mathbf{I} : \mathbf{A} = A_{ii}$. Herein the summation convention over repeated indices is employed. A superimposed dot indicates a time derivative, i.e. $\dot{\mathbf{A}} = d\mathbf{A}/dt$, and the symbol $[\![\mathbf{A}]\!]$ denotes the jump of the field quantity $\mathbf{A}$ immediately on the plus side and on the minus side of a discontinuity, i.e. $[\![\mathbf{A}]\!] = \mathbf{A}^+ - \mathbf{A}^-$.

2 THE HYPOPLASTIC CONSTITUTIVE MODEL

2.1 Inherently isotropic material

In hypoplasticity anelastic and initially inherent isotropic material properties are modeled with a constitutive equation of the rate type where the objective stress rate $\overset{\circ}{\mathbf{T}}$ is expressed by an isotropic tensor-valued function consisting of the sum of the tensor function $\mathcal{A} : \mathbf{D}$, which is linear in the rate of deformation $\mathbf{D}$, and the tensor function $\mathbf{B}\,\sqrt{\mathbf{D} : \mathbf{D}}$, which is non-linear in $\mathbf{D}$, i.e.

$$\overset{\circ}{\mathbf{T}} = \mathcal{A} : \mathbf{D} + \mathbf{B}\,\sqrt{\mathbf{D} : \mathbf{D}}. \tag{1}$$

Herein the fourth order tensor $\mathcal{A}$ and the second order tensor $\mathbf{B}$ are functions of the current Cauchy stress tensor $\mathbf{T}$ and may also depend on additional state quantities like the current void ratio e. The constitutive equation (1) is positively homogeneous of order one in $\mathbf{D}$, thus the material behaviour to be described is rate independent. With the nonlinearity in

D an anelastic material behaviour is modeled in hypoplasticity with a single constitutive equation and there is no need to distinguish between elastic and plastic parts of the deformation (Kolymbas 1988, 1991). Limit states are included in Eq.(1) for a vanishing of the stress rate (Chambon 1989, Wu & Kolymbas 1990, 2000). In order to model the influence of density and pressure on the incremental stiffness a pressure dependent relative density was introduced as an additional state variable in hypoplasticity by Wu & Bauer (1992), Bauer & Wu (1994), Wu et al. (1996). The efforts by Bauer (1995, 1996, 2000) and Gudehus (1996) led to a consistent description of the so-called SOM-states, which are characterized by a sweeping out of the memory of the loading history (Gudehus et al. 1977, Gudehus 1997). Furthermore, a factorized decomposition of equation (1) permits an easy separation and determination of the constitutive parameters and can be written as:

$$\overset{\circ}{\mathbf{T}} = f_s(e,p)\left[\mathcal{L}(\hat{\mathbf{T}}):\mathbf{D} + f_d(e,p)\,\mathbf{N}(\hat{\mathbf{T}})\sqrt{\mathbf{D}:\mathbf{D}}\right], \tag{2}$$

with:

$$\mathcal{L}(\hat{\mathbf{T}}) = \hat{a}^2\,\mathcal{I} + \hat{\mathbf{T}}\otimes\hat{\mathbf{T}}, \tag{3}$$

and

$$\mathbf{N}(\hat{\mathbf{T}}) = \hat{a}\,(\hat{\mathbf{T}} + \hat{\mathbf{T}}^*)\,. \tag{4}$$

Herein the tensors of $\mathcal{L}(\hat{\mathbf{T}})$ and $\mathbf{N}(\hat{\mathbf{T}})$ are functions of the normalized stress tensor $\hat{\mathbf{T}} = \mathbf{T}/(\mathbf{I}:\mathbf{T})$ and the deviatoric part $\hat{\mathbf{T}}^* = \hat{\mathbf{T}} - \mathbf{I}/3$. Factor $\hat{a}$ in Eq. (3) and (4) depends on the normalized stress deviator $\hat{\mathbf{T}}^*$ and the critical friction angle φ:

$$\hat{a} = \frac{\sin\varphi}{3 - \sin\varphi}\left[\sqrt{\frac{(8/3) - 3\,(\hat{\mathbf{T}}^*:\hat{\mathbf{T}}^*) + \sqrt{3/2}\,(\hat{\mathbf{T}}^*:\hat{\mathbf{T}}^*)^{3/2}\cos(3\,\theta)}{1 + \sqrt{3/2}\,(\hat{\mathbf{T}}^*:\hat{\mathbf{T}}^*)^{1/2}\cos(3\,\theta)}} - \sqrt{\hat{\mathbf{T}}^*:\hat{\mathbf{T}}^*}\right], \tag{5}$$

with the Lode-angle θ, which is defined as:

$$\cos(3\,\theta) = -\sqrt{6}\,\frac{\mathbf{I}:\hat{\mathbf{T}}^{*\,3}}{[\mathbf{I}:\hat{\mathbf{T}}^{*\,2}]^{3/2}}\,.$$

For stationary states or so-called critical states $(\mathbf{T}_c, e_c)$, which are defined for a simultaneous vanishing of the stress rate, $\overset{\circ}{\mathbf{T}} = \mathbf{0}$, and the volume strain rate, $\mathbf{I}:\mathbf{D} = 0$, factor $\hat{a}$ is equal to the Euclidean norm of the normalized stress deviator, i.e.

$$\hat{a}(\hat{\mathbf{T}}_c^*) = \hat{a}_c = \sqrt{\hat{\mathbf{T}}_c^*:\hat{\mathbf{T}}_c^*}\,. \tag{6}$$

In particular for stationary states relation $\hat{a}$ in (5) represents the stress limit condition given by Matsuoka & Nakai (1977) as shown in Figure 1a. The influence of the mean pressure $p = -\mathbf{I}:\mathbf{T}/3$ and the current void ratio e on the response of the constitutive equation (2) is taken into account by the density factor f_d, i.e.

$$f_d = \left(\frac{e - e_d}{e_c - e_d}\right)^\alpha, \tag{7}$$

and the stiffness factor f_s, i.e.

$$f_s = \left(\frac{e_i}{e}\right)^\beta\frac{1 + e_i}{e_i}\frac{h_s}{n\,h_i\,(\hat{\mathbf{T}}:\hat{\mathbf{T}})}\left(\frac{3\,p}{h_s}\right)^{1-n}, \tag{8}$$

with:

$$h_i = \frac{8\sin^2\varphi}{(3 - \sin\varphi)^2} + 1 - \frac{2\sqrt{2}\sin\varphi}{3 - \sin\varphi}\left(\frac{e_{io} - e_{do}}{e_{co} - e_{do}}\right)^\alpha\,.$$

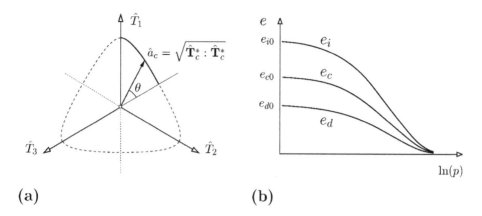

(a)   (b)

Figure 1:   (a)   Contour of the stress limit condition in the π-plane.
            (b)   Decrease of the maximum void ratio e_i, the critical void ratio e_c and the
                   minimum void ratio e_d with an increase of the mean pressure p.

Herein $\alpha < 0.5$ and $\beta > 1$ are constitutive constants. The maximum void ratio e_i, the minimum void ratio e_d and the critical void ratio e_c are pressure dependent according to

$$\frac{e_i}{e_{io}} = \frac{e_d}{e_{do}} = \frac{e_c}{e_{co}} = \exp\left[-\left(\frac{3p}{h_s}\right)^n\right],\tag{9}$$

where e_{io}, e_{do} and e_{co} are the corresponding values for $p \approx 0$ as shown in Figure 1b. In relation (9) the constant h_s with the dimension of stress scales the mean pressure p while the dimensionless exponent n reflects the degradation of the limit void ratios and the critical void ratio with increasing pressure. If the volume change of grains can be neglected, then the evolution equation for the current void ratio reads:

$$\dot{e} = (1+e)\,\mathbf{I} : \mathbf{D}\,,\tag{10}$$

which can be derived from the balance equation of mass.

2.2   Initially transversely isotropic material

In order to take into account anisotropic properties, the state variables of the hypoplastic constitutive model is extended by a structure tensor $\mathbf{S} = \mathbf{s} \otimes \mathbf{s}$ represented by the dyadic product of the normal vector $\mathbf{s}$ of the isotropic plane (bedding plane). It was proposed by Wu (1998) to incorporate the structure tensor $\mathbf{S}$ in the non-linear part of the constitutive equation for the stress rate by replacing tensor $\mathbf{N}(\hat{\mathbf{T}})$ with a symmetric second order tensor $\mathbf{M} = \mathbf{M}(\mathbf{S}, \mathbf{N}(\hat{\mathbf{T}}))$. The tensor $\mathbf{M}$ is obtained by a linear transformation of tensor $\mathbf{N}(\hat{\mathbf{T}})$ with the fourth order tensor $\mathcal{P}(\mathbf{S})$, i.e. $\mathbf{M} = \mathcal{P}(\mathbf{S}) : \mathbf{N}(\hat{\mathbf{T}})$. Thus, the extended evolution equation for the stress tensor reads:

$$\overset{\circ}{\mathbf{T}} = f_s(e, p)\left[\mathcal{L}(\hat{\mathbf{T}}) : \mathbf{D} + f_d(e, p)\,\mathbf{M}(\mathbf{S},\,\mathbf{N}(\hat{\mathbf{T}}))\,\sqrt{\mathbf{D} : \mathbf{D}}\,\right].\tag{11}$$

To further specify the form of $\mathbf{M}$ the general representation theorem for a tensor function of the two symmetric second order tensors $\mathbf{S}$ and $\mathbf{N}$ can be used (e.g. Wang 1970, Smith 1971). The selection of suitable terms from the general representation theorem is restricted by the requirement that function $\mathbf{M}$ must be linear in $\mathbf{N}$. With the additional restriction to three constants the following expression will be considered (Wu 1998):

$$\mathbf{M} = (\eta_1 + \eta_3 - 2\,\eta_2)(\mathbf{S} : \mathbf{N})\,\mathbf{S} + \eta_3\mathbf{N} + (\eta_2 - \eta_3)(\mathbf{SN} + \mathbf{NS})\,,\tag{12}$$

where η_i $(i = 1, 2, 3)$ are material parameters. It should be noted that the choice of representation (12) is motivated by a similar form originally proposed by Boehler & Sawczuk (1977) to describe the plastic behaviour of a transversely isotropic material in a simplified manner. Anisotropy is more pronounced for $\eta_i < 1$ and vanishes for $\eta_i = 1$. For the latter case the tensor function $\mathbf{M}$ in Eq.(12) reduces to $\mathbf{M} = \mathbf{N}$. It is easy to examine that with the extended constitutive equation (11) and for $\eta_i \neq 1$ the stress limit condition (6) is no longer valid, i.e. the stress ratio in a limit state is influenced by the orientation $\mathbf{s}$ of the isotropic plane (Wu 1998). In order to reproduce the experimental findings that for large monotonic shearing the stress ratio in a stationary state is independent of the initial orientation $\mathbf{s}$, an evolution of η_i corresponding to the evolution of the density factor f_d was proposed by Bauer & Huang (1999). Herein a degradation of the effect of the initial anisotropy is motivated by the experimental finding that under shearing accompanied by dilatancy, a reorientation of particles and consequently a reorientation of the direction of the contact planes takes place. Therefore, a relation between the evolution of dilatancy and the evolution of the parameter of anisotropy seems to be appropriate. In particular the rate of η_i is assumed to be proportional to the rate of f_d viz.

$$\dot{\eta_i} = \eta_{i0}\, \eta_i\, \dot{f_d}, \tag{13}$$

where η_{i0} (i=1,2,3) are constitutive constants. The integration of (13) with respect to $\eta_i(f_d = 1) = 1$ leads to

$$\eta_i = \exp\left[\eta_{i0}\left(f_d - 1\right)\right]. \tag{14}$$

It is easy to prove that for the case where the current void ratio is equal to the critical one, i.e. $e = e_c$, the value of the density factor and the parameters of anisotropy become $f_d = \eta_i = 1$. For such a critical state the corresponding critical void ratio and the critical stress ratio is independent of the initial anisotropy and the initial void ratio. The vanishing of the initial anisotropy for SOM-states also has the advantage of an easier calibration of the constitutive constants.

Altogether the hypoplastic model for an initially transverse isotropic material behaviour includes 11 constants. The values $\varphi = 30°$, $h_s = 190\,\mathrm{MPa}$, $n = 0.4$, $e_{io} = 1.20$, $e_{co} = 0.82$, $e_{do} = 0.51$, $\alpha = 0.14$, $\beta = 1.05$, $\eta_{10} = 0.8$, $\eta_{20} = 0.15$ and $\eta_{30} = 0$ are adapted to a medium quartz sand and used for the numerical calculations in Section 3.1.

3 LOCALIZATION CRITERION

In the following the possibility of a spontaneous formation of a shear band is studied for a certain state $(\mathbf{T}, e)$ and a certain bedding angle Θ with respect to a fixed Cartesian co-ordinate system as sketched in Figure 2. The shear plane or so-called discontinuity plane is characterized by a different velocity gradient $\nabla \mathbf{v}$ on either side of this plane. The jump of the velocity gradient, i.e.

$$[\![\nabla \mathbf{v}]\!] = \mathbf{g} \otimes \mathbf{n} \neq \mathbf{0}, \tag{15}$$

can be represented by the dyadic product of the unit normal $\mathbf{n}$ of the discontinuity plane and the vector $\mathbf{g}$ defining the discontinuity mode of the velocity gradient. Continuous equilibrium across the discontinuity requires (Rice & Rudnicki 1980):

$$[\![\dot{\mathbf{T}}]\!]\,\mathbf{n} = \mathbf{0}. \tag{16}$$

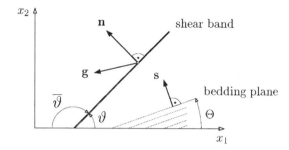

Figure 2: Orientation of the bedding plane and of the shear band.

Herein the jump of the stress rate can be related to the jump of the Jaumann stress rate, i.e.

$$[\dot{\mathbf{T}}] = [\overset{\circ}{\mathbf{T}}] + [\mathbf{W}]\mathbf{T} - \mathbf{T}[\mathbf{W}],$$

where $\overset{\circ}{\mathbf{T}}$ is the response of the hypoplastic model (11) and $\mathbf{W}$ denotes the antisymmetric part of the velocity gradient. Inserting the Jaumann stress rate into Eq.(16) leads to the relation:

$$f_s\,(\mathcal{L}:[\mathbf{D}])\,\mathbf{n} + \lambda\,f_s\,f_d\,\mathbf{M}\,\mathbf{n} + [\mathbf{W}]\,\mathbf{T}\,\mathbf{n} - \mathbf{T}[\mathbf{W}]\,\mathbf{n} = 0, \tag{17}$$

with:

$$[\mathbf{D}] = \frac{1}{2}\left[\mathbf{g}\otimes\mathbf{n} + \mathbf{n}\otimes\mathbf{g}\right],$$

$$[\mathbf{W}] = \frac{1}{2}\left[\mathbf{g}\otimes\mathbf{n} - \mathbf{n}\otimes\mathbf{g}\right],$$

and

$$\lambda = [\sqrt{\mathbf{D}:\mathbf{D}}].$$

At the onset of shear banding the stress and the void ratio are the same on either side of the discontinuity plane. Thus, the quantities f_s, f_d, $\mathcal{L}$ and $\mathbf{M}$ are also the same and they are independent of the velocity gradient. It is a peculiarity in hypoplasticity that the possibility of different incremental stiffnesses due to a different velocity gradient on either side of the discontinuity is taken into account by the single relation (17) and there is no need to distinguish whether the material outside the shear band undergoes loading or unloading (e.g. Chambon & Desrues 1985, Wu & Sikora 1991, Bauer & Huang 1997). Relation (17) can be rewritten as:

$$\mathbf{K}\,\mathbf{g} = \lambda\,\mathbf{r},$$

or

$$\mathbf{g} = \lambda\,\mathbf{K}^{-1}\,\mathbf{r}, \tag{18}$$

with:

$$\mathbf{K} = f_s\left[\hat{a}^2(\mathbf{I} + \mathbf{n}\otimes\mathbf{n})\frac{1}{2} + (\hat{\mathbf{T}}\,(\mathbf{n}\otimes\mathbf{n}))\,\hat{\mathbf{T}}\right] + \frac{1}{2}\left[(\mathbf{n}\,(\mathbf{T}\mathbf{n}))\,\mathbf{I} - (\mathbf{n}\otimes\mathbf{n})\mathbf{T} - \mathbf{T} + \mathbf{T}(\mathbf{n}\otimes\mathbf{n})\right],$$

and

$$\mathbf{r} = -f_s\,f_d\,\mathbf{M}\,\mathbf{n}\,.$$

Inserting relation (18) for $\mathbf{g}$ into the norm of $[\![\,\mathbf{D}\,]\!]$, i.e.

$$\sqrt{[\![\,\mathbf{D}\,]\!] : [\![\,\mathbf{D}\,]\!]} = \sqrt{\frac{1}{2}\Big[\mathbf{g}\,\mathbf{g} + (\mathbf{g}\,\mathbf{n})^2\Big]} = \gamma \tag{19}$$

leads to the bifurcation condition:

$$f(\vartheta) = \sqrt{\frac{(\mathbf{K}^{-1}\mathbf{r})\,(\mathbf{K}^{-1}\mathbf{r}) + ((\mathbf{K}^{-1}\mathbf{r})\,\mathbf{n})^2}{2}} - \frac{\gamma}{|\lambda|} = 0\,. \tag{20}$$

The components of unit normal $\mathbf{n}$ of the discontinuity plane are related to the unknown shear band inclination angle ϑ, i.e. $\mathbf{n} = [\,-\sin\vartheta,\ \cos\vartheta\,]^T$ with respect to the co-ordinate system in Figure 2. $\mathbf{K}$ and $\mathbf{r}$ depend on the current state quantities e, $\mathbf{T}$ and on the inclination angle Θ of the bedding plane. In order to find the lowest possible bifurcation stress ratio the value of $\gamma/|\lambda|$ can be set equal to 1 (Wu & Sikora 1992, Bauer 1999). Thus, relation (20) represents an equation for the unknown ϑ.

3.1 Bifurcation analysis

In the following the bifurcation condition (20) is examined for stress paths which are related to homogeneous compression under plane strain conditions and a constant lateral pressure starting from an isotropic stress state. Attention is paid to the lowest stress ratio where a shear band bifurcation is possible. Moreover, the influence of the the initial void ratio and the bedding angle on the bifurcation stress and the corresponding shear band inclination is also studied. All calculations are based on the single set of constitutive constants given in Section 2.

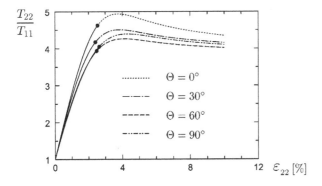

Figure 3: Stress-strain relation and onset of shear band bifurcation for homogeneous compression under plane strain conditions and a constant lateral pressure $T_{11} = -100$ kPa, an initial void ratio $e_0 = 0.6$ and different bedding angles Θ.

Figure 3 shows predictions of the stress-strain relation for different bedding angles. For homogeneous compression the maximum stress ratio is higher for $\Theta = 0°$ and becomes a minimum for $\Theta \approx 60°$, which is in accordance with experimental observations (e.g. Lam & Tatsuoka 1988). After the peak the stress ratio decreases and tends towards a

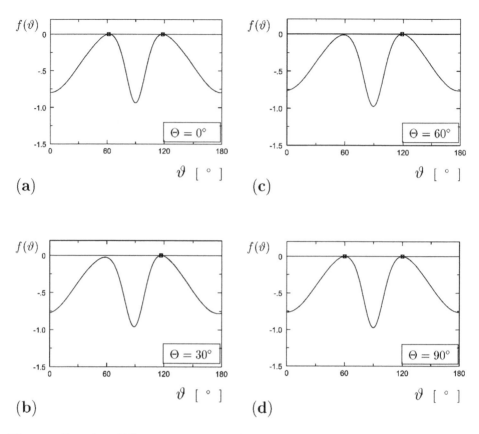

Figure 4: Function $f(\vartheta)$ corresponds to bifurcation point for bedding angle (a) $\Theta = 0$; (b) $\Theta = 30°$; (c) $\Theta = 60°$; (d) $\Theta = 90°$.

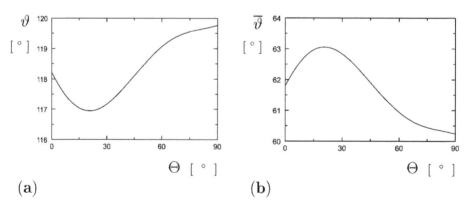

Figure 5: (a) Inclination angle ϑ versus bedding angle Θ and (b) supplementary inclination angle $\bar{\vartheta} = 180° - \vartheta$ versus bedding angle Θ.

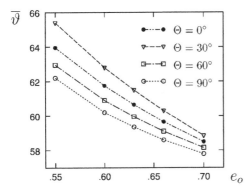

Figure 6: Shear band inclination $\overline{\vartheta}$ versus initial void ratio e_o.

stationary value which is independent of the bedding angle. The lowest bifurcation stress ratio obtained from the bifurcation analysis occurs before the peak state and it is marked by a bigger dot. Therefore the solid curves in Figure 3 denote states in which a spontaneous shear band bifurcation is not possible. But states above the first bifurcation point (dotted /dashed curves) again fulfill criterion (20) as discussed in detail for an inherently isotropic material by Bauer (1999). The lowest bifurcation stress ratio is higher for a lower bedding angle. Solutions of the bifurcation condition (20) are analysed within $0° \leq \vartheta \leq 180°$ as shown in Figure 4 for bedding angles of $0°$, $30°$, $60°$ and $90°$. The analysis shows that for $\Theta = 0°$ and $\Theta = 90°$ (Figure 4a, 4d) two shear band inclination angles ϑ are obtained at the same bifurcation stress state for $f(\vartheta) = 0$. The two possible shear bands are symmetric to the x_2-axis. However, within the range of $0° < \Theta < 90°$ only one single shear band is possible at the lowest bifurcation stress state and the corresponding shear band inclination angles ϑ are greater than $90°$ as shown in Figure 5a. The predicted shear band inclination has a minimum for $\Theta \approx 20°$ while the maximum value is obtained for $\Theta = 90°$. The supplementary inclination angle $\overline{\vartheta} = 180° - \vartheta$ versus the bedding angle Θ is represented in Figure 5b. The influence of the initial void ratio e_0 on the shear band inclination $\overline{\vartheta}$ is shown for various bedding angles Θ in Figure 6. The shear band inclination is higher for a lower void ratio and decreases with an increase of the void ratio. The result in Figure 4b and in Figure 4c shows that the difference between the two peak values for $f(\vartheta)$ is very small, which indicates that a shear band within the range of $0° < \vartheta < 90°$ could also appear as a result of deviations from the ideal conditions assumed for the present investigation. Moreover, for states beyond the first bifurcation point more than one shear band is possible and also shear bands within the range of $0° < \vartheta < 90°$ will occur. In contrast to an isotropic material the inclination of the second shear band obtained for an anisotropic material is usually not symmetric to the x_2 axis as can be detected in Figure 4b, 4c.

It should be mentioned that the present bifurcation analysis does not reflect the complete bifurcation and stability behaviour of a standard biaxial compression test with a mixed control of the stress and displacement boundary conditions. For an anisotropic material the principal stresses and the principal strains are not coaxial so that the stress and void ratio distribution with a real specimen becomes inhomogeneous as a result of the off-axes loading except for bedding angles of $0°$ and $90°$. Finite element calculations with the corresponding boundary conditions for standard biaxial compression tests and for inclination angles of the bedding plane within $0° < \Theta < 90°$ lead to inclination angles for the shear bands within the same range of $0° < \vartheta < 90°$, which shows the influence of off-axes loading (Huang 2000). The present predictions concern the initiation of a shear band depending on the stress state, the void ratio and the orientation of the bedding plane, which is also of importance in the course of finite element analyses to detect the possibility of a shear band bifurcation as a precursor of failure. The further evolution of a shear band is strongly influenced by the entire boundary value problem and non-local effects, i.e. the shear band thickness is related to a certain characteristic material length, which cannot be captured with a classical continuum model (e.g. Mühlhaus & Vardoulakis 1987, Tejchman & Gudehus 2001, Huang et al. 2002). Moreover, the shear band analysis can also be carried out for general states of stress and strain in similar way as shown for an isotropic material by Wu (2000). In contrast to an isotropic material, however, the solutions obtained for the shear band inclination will usually lose their symmetry due to anisotropy.

4 CONCLUSIONS

A hypoplastic constitutive model for a cohesionless and initially transversely isotropic granular material has been represented. The constitutive model covers a broad range of densities and pressures using only one set of constitutive constants. With an evolution equation for the parameters of transverse isotropy the influence of anisotropy declines for large shearing and the stress ratio and the volume strain tend towards a stationary state which is independent of the initial state. The influence of initially transversely isotropic material behaviour on the possibility of a spontaneous shear band bifurcation has been investigated for stress paths related to plan strain compression. The analysis shows that shear banding is strongly influenced by the stress state, the current void ratio and the orientation of the bedding plane. The lowest bifurcation stress ratio occurs before the peak and it is higher for a lower bedding angle. At the lowest bifurcation stress ratio only a single shear band is possible with the exception of loading parallel or perpendicular to the direction of the bedding plane where two symmetric shear bands appear. The inclination of the shear band is higher for lower void ratios and lower bedding angles.

REFERENCES

Arthur, J. R. F & Phillips, A. B. 1975. Homogeneous and Layered Sand in Triaxial Compression. *Géotechnique*, Vol. 25, No. 4, 799–815.

Bauer, E. & Wu, W. 1994. Extension of Hypoplastic Constitutive Model with Respect to Cohesive Powders. *Proceed. of the Eighth Intern. Conf. on Computer Methods and Advances in Geomechanics*, IACMAG 94, Siriwardane & Zaman (eds), Balkema press, 531–536.

Bauer, E. 1995. Constitutive Modelling of Critical States in Hypoplasticity. *Proceedings of the Fifth International Symposium on Numerical Models in Geomechanics*, NUMOG V, Pande & Pietruszczak (eds), Balkema press, 15-20.

Bauer, E. 1996. Calibration of a comprehensive hypoplastic model for granular materials. *Soils and Foundations*, Vol. 36, No. 1, 13–26.

Bauer, E. & Huang, W. 1997. The dependence of shear banding on pressure and density in hypoplasticity. *Proc. of the 4th Int. Workshop on Localisation and Bifurcation Theory for Soils and Rocks*, Adachi, Oka & Yashima (eds), Balkema, 1998, 81–90.

Bauer, E. & Huang, W. 1999. Effect of initial anisotropy on shear banding in granular materials. *Proc. of the 7th Int. Symposium on Numerical Models in Geomechanics*, NUMOG VII, Pande, Pietruszczak & Schweiger (eds), Balkema press, 121–126.

Bauer, E. 1999. Analysis of shear band bifurcation with a hypoplastic model for a pressure and density sensitive granular material. *Mechanics of Materials*, 31, 597–609.

Bauer, E. & Huang, W. 1999. Numerical study of polar effects in shear zones. *Proc. of the 7th Int. Symp. on Num. Models in Geomechanics*, NUMOG VII, Pande, Pietruszczak & Schweiger (eds), Balkema press, 133–138.

Bauer, E. 2000. Conditions for embedding Casagrande's critical states into hypoplasticity. *Mechanics of Cohesive-Frictional Materials*, 5, 125–148.

Boehler, J. P. & Sawczuk, A. 1977. On Yielding of Oriented Solids. *Acta Mechanica*, Vol. 27, 185–206.

Chambon, R. & Desrues, J. 1985. Bifurcation par localisation et non linéarité incrémentale: un exemple heuristique d'analyse complète. *Plastic Instability*, Presses de l'ENPC ed., Paris, 101–113.

Chambon, R. 1989. Une classe de lois de comportement incrémentalement non-linéaires pour les sols non visqueux, résolution de quelques problems de cohérence. *C.R. Acad. Sci., Paris*, 308 (II), 1571–1576.

Darve, F. 1991. Incrementally non-linear constitutive relationships. Geomaterials: Constitutive Equations and Modelling, Darve (ed), Elsevier press, 213–237.

Gudehus, G., Goldscheider, M. & Winter, H. 1977. Mechanical Properties of Sand and Clay and Numerical Integration Methods: Some Sources of Errors and Bounds of Accuracy. *In Finite Elements of Geomechanics*, Gudehus (ed), John Wiley, New York, 121–150.

Gudehus, G. 1996. A comprehensive constitutive equation for granular materials. *Soils and Foundations*, Vol. 36, No. 1, 1–12.

Gudehus, G. 1997. Attractors, percolation thresholds and phase limits of granular soils. *Powders and grains*, Behringer & Jenkins (eds), Balkema, 169-183.

Hill, R. J. 1962. Acceleration waves in solids. *J. Mech. Phys. Solids*, No. 10, 1–16.

Huang, W. 2000. Hypoplastic modelling of shear localisation in granular materials. Doctoral Thesis, Graz University of Technology.

Huang, W., Nübel, K. & Bauer, E. 2002. Polar extension of a hypoplastic model for granular materials with shear localization. *Mechanics of Materials*, 34, 563–576.

Kolymbas, D. 1988. Eine konstitutive Theorie für Böden und andere körnige Stoffe. *Publication Series of the Institute of Soil Mechanics and Rock Mechanics, Karlsruhe University*, No. 109.

Kolymbas, D. 1991. An outline of hypoplasticity. *Arch. of Appl. Mechanics*, 3, 143–151.

Lam, W. K. & Tatsuoka, F. 1988. Effect of Initial Anisotropic Fabric and σ_2 on Strength and Deformation Characteristics of Sand. *Soils and Foundations*, Vol. 28, No. 1, 89–106.

Matsuoka, H. & Nakai, T. 1977. Stress-strain relationship of soil based on the 'SMP'. *Proc. of Speciality Session 9, IX Int. Conf. Soil Mech. Found. Eng., Tokyo*, 153–162.

Mühlhaus, H. B. & Vardoulakis, I. 1987. The thickness of shear bands in granular materials. *Geotechnique*, Vol. 37, 271–283.

Oda, M., Nemat-Nasser, S. & Konishi, J. 1985. Stress Induced Anisotropy in Granular Masses. *Soil and Foundations*, Vol. 25, No. 3, 85–97.

Rice, J. & Rudnicki, J. W. 1980. A note on some features on the theory of localization of deformation. *Int. J. Solids Structures*, 16, 597–605.

Rudnicki, J. W. & Rice, J. 1975. Conditions for the localization of deformation in pressure sensitive dilatant materials. *J. Mech. Phys. Solids*, Vol. 23, 371–394.

Tamagnini, C., Viggiani, G., Chambon, R. 2000. A review of two different approaches to hypoplasticity. *Constitutive Modelling of Granular Materials*, Kolymbas (ed), Springer press, 107–145.

Tatsuoka, F., Nakamura, S., Huang, C. & Tani, K. 1990. Strength anistropy and shear band direction in plane strain tests of sand. *Soils and Foundations*, Vol. 30, No. 1, 35–54.

Tejchman, J. & Gudehus, G. 2001. Shearing of a narrow granular layer with polar quantities. *Int. J. Numer. Meth. Geomech.*, 25, 1–28.

Smith, G.F. 1971. On isotropic functions of symmetric tensors, skew-symmetric tensors and vectors. *Int. J. Engng. Sci.*, 9, 899–916.

Wang, C.C. 1970. A new representation theorem for isotropic functions, parts I and II. *Archive for Rational Mechanics and Analysis*, 36, 166–223.

Wu, W., Kolymbas, D. 1990. Numerical testing of the stability criterion for hypoplastic constitutive equations. *Mechanics of Materials*, 9, 245–253.

Wu, W. & Sikora, Z. 1991. Localized bifurcation in hypoplasticity. *Int. J. Engng. Sci.*, Vol. 29, No. 2, 195–201.

Wu, W. & Bauer, E., 1992. A hypoplastic model for barotropy and pyknotropy of granular soils. *Proc. of the Int. Workshop on Modern Approaches to Plasticity*, Kolymbas (ed), Elsevier 1993, 225 – 245.

Wu, W. & Sikora, Z. 1992. Localized bifurcation in pressure sensitive dilatant granular materials. *Mech. Research Communications*, 19, 289–299.

Wu, W., Bauer, E. & Kolymbas, D., 1996. Hypoplastic constitutive model with critical state for granular materials. *Mechanics of Materials*, 23, 45–69.

Wu, W. 1998. Rational Approach to Anisotropy of Sand. *Int. J. Numer. Anal. Meth. Geomech.*, Vol. 22, 921–940.

Wu, W. & Kolymbas, D. 2000. Hypoplasticity then and now. *Constitutive Modelling of Granular Materials*, Kolymbas (ed), Springer press, 57–105.

Wu, W. 2000. Nonlinear analysis of shear band formation in sand. *Int. J. Num. Anal. Meth. Geomech.*, 24, 245–263.

Yamada, Y. & Ishihara, K. 1979. Anisotropic Deformation Characteristics of Sand under Three Dimensional Stress Conditions. *Soils and Foundations* Vol. 19, No. 2, 79–94.

4. Experimental studies

Bifurcations & Instabilities in Geomechanics, – Labuz & Drescher (eds.)
© 2003 Swets & Zeitlinger, Lisse, ISBN 90 5809 563 0

Recent progress in experimental studies on instability of granular soil

J. Chu
Nanyang Technological University, Singapore (cjchu@ntu.edu.sg)

W. K. Leong
The University of Auckland, New Zealand (l.kai@auckland.ac.nz)

ABSTRACT: A review of some case studies indicates that granular soil can become unstable under a completely drained condition and instability can also occur for dilative sand. A laboratory study on the instability of dilative sand and the instability of sand under drained conditions was carried out. Some experimental data are presented. Classification of instability of sand into runaway and conditional instability is made. The differences between the two types of instability and the factors affecting the different types of instability are discussed. The practical implications of each type of instability are discussed.

1 INTRODUCTION

Liquefaction and instability behaviour of sand have often been studied under undrained conditions mainly for loose sand (Casagrande 1975, Lade 1993, Sasitharan et al. 1993, Leong et al. 2000). Instability and static liquefaction under undrained conditions are often considered the triggering factors for the failure of loose granular soil slopes. However, there are cases where instability occurred under essentially drained conditions. In a recent re-analysis of the Wachusett Dam failure in 1907, Olson et al. (2000) concluded that the failure was mainly triggered by static liquefaction that occurred under completely drained conditions. Through laboratory model tests, Eckersley (1990) observed that the pore water pressure increase in the gentle granular soil slope is a result, rather than the cause of flowslide. In other words, the flowslide took place under a drained condition. In fact, the limitation of studying liquefaction under undrained conditions has been recognized for a long time. Failure mechanisms related to a redistribution of void ratio within a globally undrained sand layer (Fig. 1) and spreading of excess pore pressure with global volume changes (Fig. 2) have been envisaged by National Research Council (1985) as Mechanisms B and C.

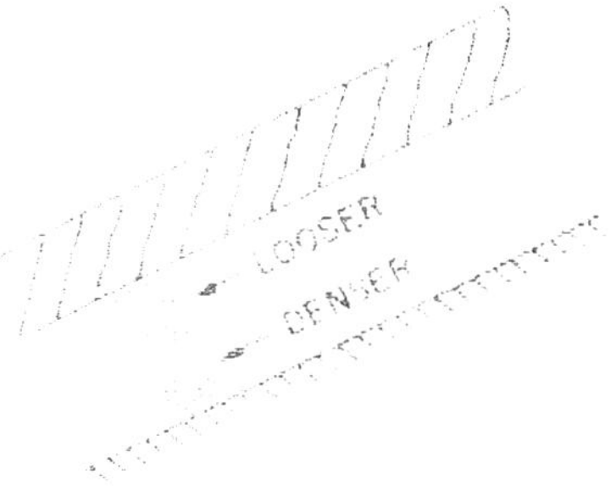

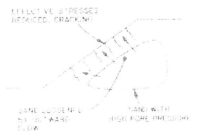

Figure 1. Mechanism B by NRC (1985): Situation for void redistribution within a globally undrained sand layer.

Figure 2. Mechanism C by NRC (1985): Situation for failure by spreading of excess pore pressure with global volume changes.

The possibility of dilating behaviour of soil masses prior to slope collapse is also observed in several case studies. Terzaghi & Peck (1967) once stated "a clean sand deposited underwater is stable, although it may be loose, because the grains roll down into stable positions. In a sand capable of spontaneous liquefaction some agent must interfere with this process". Using to-day's terminology, the above statement has been interpreted by Been et al. (1988) as "a clean fill deposited underwater will be in a dilatant rather than contractive state. To achieve a con-tractive state, there must be some agent which interferes with the deposition process." Been et al. (1988) also argued that the Nerlerk berm failure case (Sladen et al. 1985b) might have oc-curred for dilatant sand which lies below the steady state line (Been et al. 1987, Been et al. 1988). Several other cases of flow slide failure in dilatant sand have also been presented by Been et al. (1988). Casagrande (1975) indicated that prior to liquefaction and flow of large masses of rather dense granular talus in the Alps, brooks emerging from the toe of the talus de-posits stop flowing. Fleming et al. (1989) reported observations of time lags between the be-ginning of landslide movements and the initiation of debris flows. Harp et al. (1990) also ob-served abrupt decreases in the pore pressure 5-50 minutes before failure in the three slides triggered by artificial subsurface irrigation.

Therefore, in addition to liquefaction, there are other types of failure mechanisms that con-trol the stability of granular slope. However, instability[1] under other than undrained conditions, such as fully drained conditions, has seldom been studied. The mechanisms of instability of di-lating sand have not be properly investigated either.

Furthermore, slope failures are often caused not only by an increase in external load, but also by a reduction in the effective mean stress which is due to, for example, water infiltration into slopes and release of lateral stress. As suggested by Brand (1981), when investigating the fail-ure mechanisms of slope, the stress-strain behaviour of the soil along stress paths that simulate water infiltration should be studied. Such stress paths may be idealized as paths with constant shear stress, but decreasing mean effective stress, or the so called constant shear-drained (CSD) tests performed under constant deviator stress (Brand 1981, Anderson & Riemer 1995).

A research program on the instability behaviour of granular soil has been carried out at the Nanyang Technological University, Singapore, in the past years. The objectives of this paper are to introduce briefly this research program and illustrate some instability behaviour of granu-lar soil that have been observed under drained conditions for both loose and dense sand. Clas-sification of instability into two types, the runaway and conditional, is made. Another objective that the experiments can serve is to show that plastic yielding, especially instability can occur along an "unloading" path. As this could not be predicted by conventional plasticity as ex-plained by Drucker & Seereeram (1987) and Jefferies (1997), the data could be useful in verify-ing the predictive ability of some constitutive models.

2 SOIL TESTED AND EXPERIMENTAL SETUP

A marine dredged sand, which has been used in large quantity for several large land reclama-tion projects in Singapore, has been studied. The basic properties of the sand are given in Ta-ble 1. The other physical properties and the drained and undrained behaviour of the sand at loose saturated state are reported in Leong et al. (2000). The loose specimens were prepared by a moist tamping method in which the sand was premixed to a moisture content of 5% and was saturated after the formation of the specimen. The dense specimens were prepared by pluviat-ing sand in water. The nominal dimensions of the specimens were 100 mm in diameter and 200 mm in height.

Triaxial machines have often been used in the past to study the instability behaviour of sand. In addition to the use of a triaxial testing system, a plane-strain apparatus (Fig. 3) and a hollow cylinder apparatus (Fig. 4) have also been used to study the instability behaviour of soil under more generalized stress conditions. A strain path testing technique (Chu & Lo, 1991, Chu et al. 1991) has been used in these tests to study how dilation affects the instability behaviour of sand. In a strain path test, the strain increment ratio $d\varepsilon_v/d\varepsilon_1$ is controlled. An undrained test is

---

[1] As defined here as a behaviour in which large plastic strains are generated rapidly due to the inability of a soil element to sustain a given stress.

a special case when $d\varepsilon_v/d\varepsilon_1 = 0$. Some other than undrained conditions can be simulated using strain path tests with $d\varepsilon_v/d\varepsilon_1 \neq 0$.

Table 1. Basic properties of the tested sand.

Mean grain size (mm)	Uniformity coefficient	Specific gravity	Maximum void ratio	Minimum void ratio	Fines content (%)	Shell content (%)
0.3 ~0.35	2.0	2.60	0.916	0.533	0.4	14

Figure 3. Plane-strain apparatus.

Figure 4. Hollow cylinder apparatus.

3 INSTABILITY BEHAVIOUR OF GRANULAR SOIL

3.1 Instability of loose sand under undrained conditions

It is well known that when a loose sand specimen is sheared along an undrained path, an effective stress path as typically shown in Figure 5 will be obtained. Point A is the peak of effective stress path. If the test is conducted under a deformation-controlled condition, strain-softening behaviour that is characterized by a reduction in deviator stress will manifest. On the other hand, if the test is conducted under a load-controlled condition, the specimen will become unstable at point A. This behaviour has often been referred to as static liquefaction. The line, which connects the top of the effective stress paths, has been called the instability line. The zone between the instability line and the failure line (the same as the critical state line for loose sand) is called the zone of potential instability which specifies the instability condition for loose sand under undrained conditions (Lade, 1993).

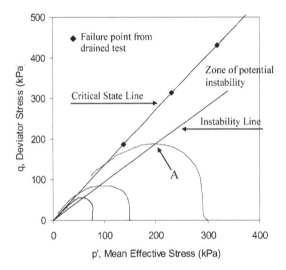

Figure 5. Zone of potential instability.

It needs to be pointed out that the instability line is not unique, but varies with the void ratio of sand and the applied effective mean stress. Figure 6 shows the effective stress paths obtained from a series of CIU tests conducted on specimens with different consolidation void ratios, e_c, but under the same mean effective stress. It can be seen from Figure 6 that the smaller the e_c, the higher the instability line. The highest instability line for loose sand is on or very close to the critical state line (CSL). It is also noted that different interpretations have been given to the instability behaviour observed by different researchers (Sladen et al. 1985b, Lade 1993, Sasitharan et al. 1993). As a result, the lines used to specify the instability conditions are different. Nevertheless, the differences among the different definitions are small, as pointed out by Lade (1993). The physical meanings behind the different interpretations are also essentially the same, i.e., to specify a yielding point where large plastic strain can develop.

Figure 7 presents the results of two instability tests, UD01 and UD02, on loose sand specimens with e_c of 0.967 for both tests. The tests were conducted by bringing the stress states to points A and C respectively along drained paths. Upon reaching points A or C, the mean total and deviator stresses started to be maintained constant and an undrained condition was imposed. The void ratios at points A and C were 0.973 and 0.938 respectively. The instability line (IL) corresponding to e = 0.972 (based on Fig. 6) is also plotted in Figure 7a. It can be seen that point A is below and point C is above the IL respectively. The axial strain and pore water pressure developments during the undrained stage are shown in Figs. 7b and 7c. It can be seen from Figure 7b that the axial strain suddenly increased at point C and instability occurred in Test UD02 but not in Test UD01. When instability occurred, the specimen physically collapsed. The excess pore water pressure also shot up, as shown in Figure 7c.

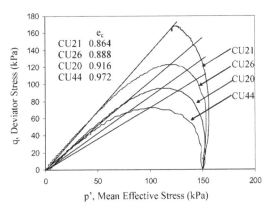

Figure 6. Non-uniqueness of instability line.

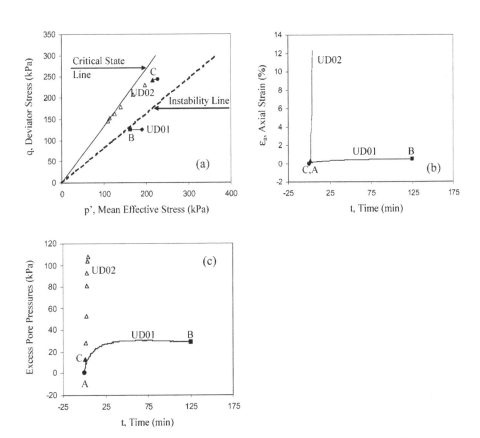

Figure 7. Instability test of loose sand under undrained condition (Tests UD01 & UD02).

It can therefore be concluded that instability will occur for loose sand under undrained conditions if the stress point is above its instability line. It has also been established that instability will not occur for loose sand under a drained condition if the stress state imposed at point C does not change (Lade & Pradel 1990, Chu 1991, and Leong et al. 2000).

3.2 Instability of loose sand under drained conditions

The instability behaviour of sand along a CSD path was also examined experimentally. The results of a typical test, DR7, are shown in Figure 8. The loose sand specimen (with a void ratio $e_c = 0.945$) was firstly sheared to point A along a drained path (Fig. 8a). The deviator stress at point A is q = 150 kPa. On the q = constant path, the confining stress was reduced at a rate of 1 kPa/min, resulting a stress path moving from point A to point B (Fig. 8a). There were little axial and volumetric strain developments until point B where both axial and volumetric strains started to develop at a faster rate, as shown in Figs. 8b and 8c. This can be seen more obviously from Figure 7d which shows that the axial strain rate shot up at point B, indicating an unstable behaviour. Using point B, the instability line can be determined as shown in Figure 8a. With further reduction in the confining stress, the stress path moved further toward the CSL. However, at this stage the axial and volumetric strain rates had increased to such an extent that the testing system could not catch up to maintain q to be constant. It needs to be pointed out that the pore water pressure did not change during the whole test (Leong 2001). Therefore, the instability in the form of a rapid increase in plastic strains is observed under a fully drained condition. Sasitharan et al. (1993) and Gajo et al. (2000) have also reported similar observations.

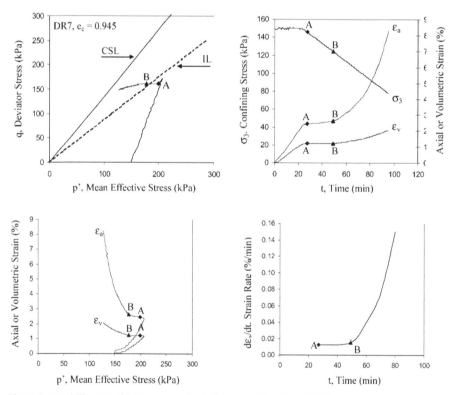

Figure 8. Instability test of loose sand under drained condition (Test DR7).

3.3 Instability of dense sand under non-undrained conditions

For dense sand, instability will not occur under undrained conditions, as concluded by Chu et al. (1993) and Chu & Leong (2001). As there is no reason why a soil has to deform under a completely undrained condition, a soil element could be subjected to other than non-undrained conditions. This non-undrained condition can be modeled experimentally by strain path with $d\varepsilon_v/d\varepsilon_1$ controlled. When $d\varepsilon_v/d\varepsilon_1 > 0$ is imposed on dense sand, the pore water pressure will decrease and instability will not occur. However, when an adequate dilative $d\varepsilon_v/d\varepsilon_1$ is imposed, the pore water pressure will increase and instability becomes a possibility.

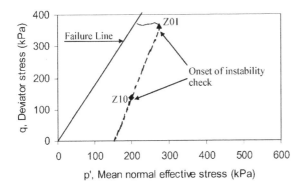

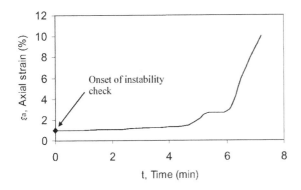

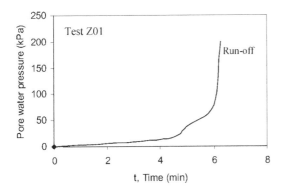

Figure 9. Instability test of dense sand under non-undrained condition (Tests Z01 & Z10).

Figure 9(a) presents the effective stress paths resulting from two stability tests, Z01 and Z10, on dense sand. In conducting the tests, the specimens were first sheared along a drained path with $\sigma_{30}' = 150$ kPa to a stress ratio of $q/p' = 1.36$ and 0.75 (or $\sigma_1'/\sigma_3' = 3.5$ and 2.0) for tests Z01 and Z10 respectively. The external loads, i.e., the axial load and the cell pressure, were then maintained constant to conduct an instability check along a strain path of $d\varepsilon_v/d\varepsilon_1 = -0.67$. Under these conditions, instability occurred for Test Z01, but not for Test Z10. When instability occurred in Test Z01, the axial strain and the pore-water pressure increased suddenly, as shown in Figs. 9(b) and 9(c). Physically it was observed that the specimen collapsed suddenly, i.e., prefailure instability had occurred.

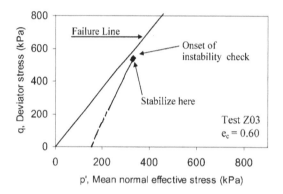

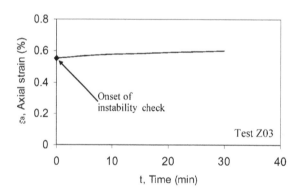

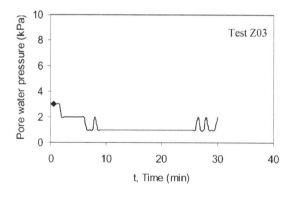

Figure 10. Instability test of dense sand under non-undrained condition (Test Z03).

182

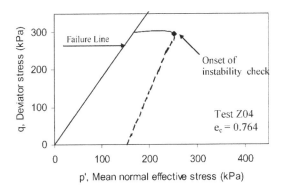

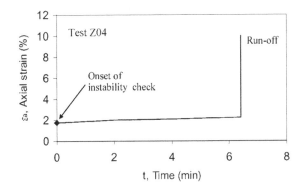

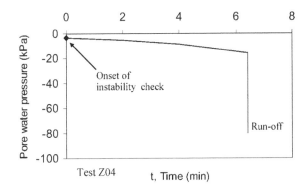

Figure 11. Instability test of dense sand under non-undrained condition (Test Z04).

Tests Z03 and Z04 were conducted by following the same path but on specimens with different densities. The specimen used for Test Z03 was dense with $e_c = 0.60$ and for Test Z04 medium loose with $e_c = 0.764$. Both tests were conducted by shearing the specimens along a drained path with $\sigma_{30}' = 150$ kPa to a stress ratio of $q/p' = 1.2$ (or $\sigma_1'/\sigma_3' = 3.0$) and then switching to a strain path of $d\varepsilon_v/d\varepsilon_1 = -0.54$. The deviator load and cell pressure were maintained constant in both tests during the strain path control. The results of Tests Z03 and Z04 are shown in Figs. 10 and 11 respectively. It can be seen that pre-failure instability occurred for Test Z04 (Fig. 11), but not for Test Z03 (Fig. 10).

It needs to be pointed out that the instability shown in Figs. 9 and 11 was not only due to the control of a negative strain increment ratio. This is testified by Tests Z10 (Fig. 9) and Z03 (Fig. 10) where instability did not occur along the same strain paths. Instability only occurs when appropriate conditions are met. The observed instability was not due to strain localization or rate/time effect either (Chu et al. 1993).

3.4 Instability of dense sand under drained conditions

It has been concluded by Chu (1991) and Leong et al. (2000) that when the stress state is maintained constant, instability will not occur for dense sand under drained conditions. However, if a dense specimen is sheared along a CSD path, instability may become a possibility.

Two triaxial tests on dense specimens, Tests DR39 and DR40, were conducted under q = 303 kPa (point A) and q = 343 kPa (point A') respectively. The void ratios of the specimens after consolidation were 0.657 for DR39 and 0.647 for DR40. The effective stress paths of the two tests are presented in Figure 12. The CSL and the failure line (FL) are also plotted in Figure 12 for reference. The axial and volumetric strains versus time and the axial and volumetric strains versus mean effective stress curves are shown in Figure 13 for Tests DR39. It can be seen from Figure 13a that the axial strain started to increase abruptly at point B for Tests DR39, signifying the onset of instability. Therefore, the line passing though Point B defines the IL. The volumetric strain in the form of dilation also began to develop rapidly at point B. It can also be seen from Figure 13b that large yielding starts to develop at point B and therefore, point B is also a yield point for Test DR39. Similarly, point B' is a yield point for Test DR40.

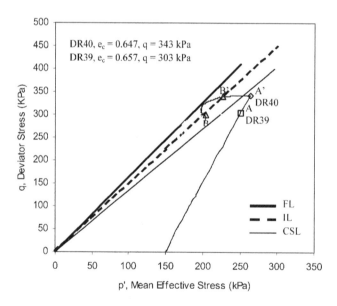

Figure 12. Instability test of dense sand under drained condition (Tests DR39 & DR40).

4 CLASSIFICATION OF INSTABILITY

The instability behaviours of both loose and dense sand occurring under undrained, non-undrained, and drained conditions are presented above. It needs to be pointed out that there are fundamental differences between the instability occurring under different conditions. In general, the instability can be classified into two major types: runaway and conditional. Instability of loose sand under undrained conditions and dense sand under non-undrained conditions are runaway type, whereas instability of sand under drained conditions are conditional type.

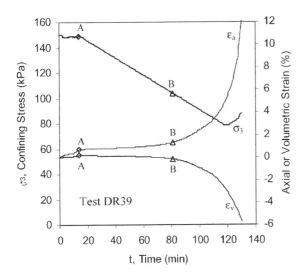

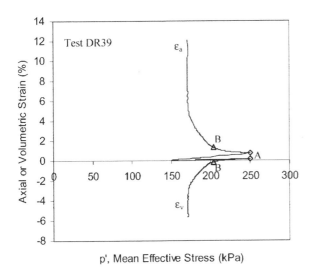

Figure 13. Instability test of dense sand under drained condition (Test DR39).

4.1 Runaway instability

Runaway instability refers to the type that once large deformation starts to develop, the specimen will collapse suddenly or within a very short time. The undrained instability of loose sand (see Fig. 7) and the instability of dense sand under non-undrained conditions (see Figs. 9 and 11) belong to this type. Both are characterized by an increase in pore water pressure during the instability. The mechanisms, which control this type of instability, are the same and can be described as follows.

It is well known that when sand of different void ratios is sheared under a drained condition, the volumetric strain behaviour will be different. If the sand is very loose, the volumetric strain will be contractive. On the other hand, if the sand is dense, it will dilate. A drained test defines the volumetric strain response of sand to a zero pore water pressure change condition. In other words, in order to maintain the pore water pressure to be constant, the volume of the sand will have to change in the way as measured in a drained test. In a drained test, loose sand needs to contract, that is to discharge water, to maintain the pore water pressure to be constant. On the other hand, when volume change, i.e., water discharge, is not allowed in an undrained test, the pore water pressure will increase. As a consequence, the effective confining stress will decrease and the shear resistance will decrease accordingly. Static liquefaction or runaway instability will occur. Similarly, in a drained test, a dense sand need to dilate, i.e., to absorb water, to maintain the pore water pressure to be constant. Under an undrained condition, water is not allowed to flow into the specimen and hence the pore water pressure inside the specimen will reduced. This is what is observed in an undrained test on dense sand. However, if an non-undrained condition is imposed to force the specimen to dilate more than the sand would under a drained condition, extra water will have to flow into the specimen to generate the required dilation. As a result, positive pore water pressure will develop. The specimen will become unstable in a similar way as in an undrained test.

The occurrence of runaway instability is affected by factors such as, the drainage conditions which can be simulated by the strain increment ratio imposed, $(d\varepsilon_v/d\varepsilon_1)_i$, the void ratio of soil, and the initial effective stress ratio. The effects of drainage condition and void ratio of soil have been explained through the comparison of the results presented in Figs. 10 & 11 for Tests Z03 and Z04. As discussed in detail by Chu et al. (1993), the instability mechanism for both loose and dense sand can be expressed by the differences between the strain increment ratio imposed, $(d\varepsilon_v/d\varepsilon_1)_i$, and the strain increment ratio of the soil, $(d\varepsilon_v/d\varepsilon_1)_s$, and instability occurs when:

$$\left(\frac{d\varepsilon_v}{d\varepsilon_1}\right)_i - \left(\frac{d\varepsilon_v}{d\varepsilon_1}\right)_s \leq 0 \tag{1}$$

The strain increment ratio of the soil, $(d\varepsilon_v/d\varepsilon_1)_s$, is the maximum $d\varepsilon_v/d\varepsilon_1$ ratio determined from the ε_v versus ε_1 curve of a drained test conducted under the same initial effective confining stress. For very loose sand, $(d\varepsilon_v/d\varepsilon_1)_s$ is greater than or equal to zero. Instability tends to occur when $(d\varepsilon_v/d\varepsilon_1)_i = 0$, i.e., under an undrained condition. For dense sand, $(d\varepsilon_v/d\varepsilon_1)_s$ is negative. Instability is only possible when $(d\varepsilon_v/d\varepsilon_1)_i$ is even more negative than $(d\varepsilon_v/d\varepsilon_1)_s$. The $(d\varepsilon_v/d\varepsilon_1)_s$ for the specimens used in Tests Z03 and Z04 were -0.50 and -0.47. This explains why instability occurred in Test Z04 when $(d\varepsilon_v/d\varepsilon_1)_i = -0.54$ was imposed in both tests.

Constitutive equations can be used to predict the value of $(d\varepsilon_v/d\varepsilon_1)_s$ in Inequality (1). If the modified Rowe's stress-dilatancy theory (Wan & Gou, 1988) is used for a drained test, $(d\varepsilon_v/d\varepsilon_1)_s$ can be calculated as:

$$\left(\frac{d\varepsilon_v}{d\varepsilon_1}\right)_s = 1 - \frac{1}{K}\left(\frac{\sigma_1'}{\sigma_3'}\right)_f \tag{2}$$

where $(\sigma_1'/\sigma_3')_f$ = failure stress ratio achieved in a drained test, and K is expressed as:

$$K = \frac{1+\sin\phi^*}{1-\sin\phi^*} \quad \text{with} \quad \sin\phi^* = (e/e_{cr})^\alpha \sin\phi_{cr} \tag{3}$$

where ϕ^* is the friction angle mobilized along a certain macroscopic plane which evolves during deformation history, ϕ_{cr} and e_{cr} are the friction angle and void ratio at critical state.

Combining (2) with Inequality (1) leads to:

$$\left(\frac{d\varepsilon_v}{d\varepsilon_1}\right)_i \leq 1 - \frac{1}{K}\left(\frac{\sigma_1'}{\sigma_3'}\right)_f \tag{4}$$

Since $(\sigma_1'/\sigma_3')_f$ depends on the initial void ratio and to a certain extent the effective confining stress, Inequality (4) reflects the influence of these two factors, in addition to $(d\varepsilon_v/d\varepsilon_1)_i$, on instability.

The effect of stress ratio is specified by the instability line as shown in Figure 6. The same instability line exists for dense sand under non-undrained conditions. A comparison of Tests Z01 and Z10 (Fig. 9) shows that the stress ratio, q/p', has to be sufficiently high to induce instability. This can be explained by plasticity theory. For plastic flow to develop, the yield surface has to expand, which requires the stress increment to point outside the yield surface. A typical yield surface for sand is shown in Figure 14. The effective stress path involved in an instability test is characterized by $dq \leq 0$ and $dp' < 0$. The deviator stress q decreases because the cross-section of the specimen increases while the deviator load is maintained constant. The effective mean stress decreases due to the increase in pore pressure. As shown in Figure 14, such a stress path only points outside the yield surface when it is within the hatched zone. In terms of stress state, this hatched zone has a stress ratio higher than the stress ratio at the peak point of the yield surface. If the stress ratio where an instability test starts is lower than the stress ratio at the peak point, such a stress path will point inside the yield surface, as shown in Figure 14, and result in an elastic response. The hatched area is characterized by a negative normal direction. Mathematically, this can be expressed as:

$$\frac{\partial f}{\partial p'} < 0 \tag{5}$$

where f is the yield surface. Inequality (5) defines the stress conditions under which instability may occur.

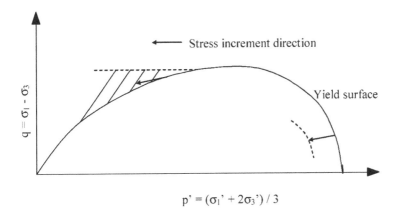

Figure 14. Yield surface and instability behaviour.

4.2 Conditional instability

For runaway instability, the pore water increases as a result of strain path control with undrained condition as a special case. Consequently, the effective mean stress reduces and instability occurs when the effective stress path points outside of the yield surface within the hatched zone.

Similar effective stress paths can be produced under a drained condition when the effective mean stress is reduced at a controlled manner. This is how drained instability occurs in a CSD test where the effective mean stress is being reduced. If there is no reduction in the effective mean stress, instability will not occur, as observed by Lade et al. (1988) and Chu et al. (1993). Therefore, the occurrence of drained instability is dependent on the reduction of the lateral stress. That is why it is a conditional instability.

4.3 Physical meanings

Instability as defined in this paper refers to a behaviour in which large plastic strains are generated *rapidly*. For large plastic strain to develop, the soil must be in a yielding state. Therefore, yielding is the necessary condition for instability. This has been explained by Lade (1992) for the instability occurring for loose sand under undrained conditions. The instability line, in fact, is defined based on the yielding conditions (Lade 1992, Chu et al, 1993). As such, the zone of instability is defined regardless of the drainage condition. This explains why instability can occur under both undrained and drained conditions as long as the stress path leads the stress state into the zone of instability. The instability line is defined as a line linking the peak points of the yield surfaces. It should be pointed out that for constant deviator stress path, yielding is not precisely defined by the instability line. However, the instability line can be used as a good approximation. This is supported by the experimental data presented in this paper and other similar tests (Leong & Chu, 2002, Chu & Leong, in press).

The instability that occurred in Tests DR39 and DR40 for dense sand was also associated with plastic yielding. It can be seen from the strains versus effective mean stress curves shown in Figure 13b that point B, where instability began, is also the yield point for Test DR39.

Although yielding is the necessary condition for instability to occur, it is not sufficient. In other words, plastic yielding does not necessarily cause a soil specimen to be unstable. Yielding means the development of a large strain for a small change in stress. It does not imply that the specimen will become unstable which is characterized by a sudden increase in the strain increment rate, $d\varepsilon_1/dt$ (see Fig. 8d). Therefore, we cannot assume that yielding is automatically the condition for instability. Whether instability can occur along a given stress path needs to be established separately.

The CSD tests conducted on dense sand have confirmed the common belief that the denser the sand the higher the stress ratio required to induce any plastic yielding or unstable behaviour. The data also implies that stability design using the peak friction angle for dense sand can be unconservative, as given the conditions even dense sand can become unstable at a stress ratio below failure.

4.4 Runaway versus Conditional Instability

Although instability is observed to occur for loose sand under both undrained and drained conditions, the two types of instability are fundamentally different. Firstly, the strain rates developed during instability are different. The average strain rates measured during instability for Tests UD2, DR7 and DR40 are compared in Figure 15. It can be seen that the specimen in Test UD2 collapsed much faster during an undrained instability (Test UD2). In fact, the specimen collapsed almost instantly. Therefore, the instability that occurred under undrained conditions is a runaway type. On the other hand, the specimen did not collapse instantly during a drained instability (Tests DR7 and DR40), although the strain rate was increasing at an accelerating rate. The differences between runaway and conditional instabilities are mainly due to the differences in the effective mean stress reduction mechanisms. In an undrained or non-undrained test, the pore water pressure increases as a response to the external loading condition. Once the instability conditions are met, the pore water pressure can increase very rapidly at an accelerating rate. Thus, the effective mean stress will also reduce very rapidly at an accelerating rate. However, in a drained test, the reduction in the effective mean stress is controlled. There is a lack of accelerating effect. As a result, the instability is not a runaway type, although instability is developing as the strains start to develop rapidly at an accelerating rate.

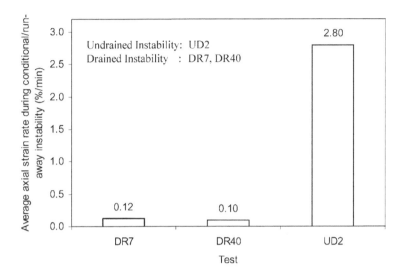

Figure 15. Average axial strain rate during drained/undrained instability.

The relationship between runaway instability and conditional instability is like the relationship between quicksand and liquefaction. Although both occur when the pore water pressure has increased to the extend that the effective stress becomes zero, the pore water pressure increase due to a control in seepage (as in quicksand) is different from the pore water pressure generation as a response to the external dynamic load (as in liquefaction).

Despite of the differences in the two types of instability, the conditions that govern the both types of instability for loose sand are the same. This is because for loose sand, both the CSD and the undrained tests produce stress paths crossing the instability line, i.e., the yield surface.

It is because the conditions for both types of instability are the same for loose sand, when a conditional instability occurs under a drained condition, it may evolve into a runaway instability if the drainage is insufficient to dissipate all the pore water pressure, e.g. during an earthquake. This was observed in some of the tests. The drained collapse behaviour reported by Sasitharan et al. (1993) may belong to this type; in which "a slight increase in pore water (6.7 kPa) followed by a catastrophic undrained failure or collapse of the sample" was observed. Similar observations were made by Eckersley (1990) from his model tests. He reported that static liquefaction had occurred for loose sand under essentially static, drained conditions (Eckersley 1990). The flow slides in Eckersley's experiments were initiated by slow water level increases, i.e., along a stress path with a reduction in effective mean stress. Therefore, the flow slides in these model tests are likely caused by drained instability. Eckersley (1990) also reported that excess pore water pressures were developed after the start of flow slides. This pore water pressure building up was likely caused by the inability to dissipate fully the pore water pressure, which was generated as a result of large volumetric strain development.

5 CONCLUSION

A review of some case studies indicates that granular soil can become unstable under a completely drained condition and instability can also occur for dilative sand. Laboratory studies have also shown that even dilative sand can become unstable under given conditions and instability can occur for sand under drained conditions. Four types of instability behaviour are presented in this paper, namely the instability of loose sand under undrained conditions, the instability of loose sand under drained conditions, instability of dense sand under non-undrained conditions, and instability of dense sand under drained conditions. The observed instability behaviour can be classified into two types, the runaway instability and the conditional instability. The two types of instability are fundamentally different. The differences between the two types of instability and the factors affecting the two types of instability are discussed. Using the observed instability data, some failure cases, which could not be interpreted by undrained liquefaction before, can now be explained. The present study is limited to axisymmetric conditions. Further study on the instability behaviour of granular soil under more generalized stress conditions is currently being studied.

6 PREFERENCES

Anderson, Scott A. & Riemer, M. F. 1995. Collapse of saturated soil due to reduction in confinement. *Journal of Geotechnical Engineering* 121(2): 216-220.

Been, K. Conlin, B. H. Crooks, J. H. A. Fitzpatrick, S. W. Jefferies, M. G. Rogers, B. T. & Shinde, S. 1988. Back analysis of the Nerlerk berm liquefaction slides:[1] Discussion. *Canadian Geotechnical Journal* 24: 170-179.

Brand, E. W. 1981. Some thoughts on rain-induced slope failures. *Proceedings of the 10th ICSMFE.* (3): 373-376. Stockholm.

Casagrande, A. 1975. Liquefaction and cyclic deformation of sands – a critical review. *Proc. 5th Pan American Conference On Soil Mechanics And Foundation Engineering.* Buenos Aires; also published as *Harvard Soil Mechanics Series No. 88,* Cambridge: Mass.

Chu, J. 1991. Strain softening behaviour of granular soils under strain path testing. *Ph.D. Thesis* University of New South Wales. Canberra. Australia.

Chu, J. & Lo, S-C. R. 1991. On the implementation of strain path testing. *Proceedings of the International Conference on Soil Mechanics and Foundation Engineering* (1): 53-56.

Chu, J. Lo, S-C. R. & Lee, I. K. 1991. Strain softening behaviour of a granular soil in strain path testing. *Journal of Geotechnical Engineering* ASCE. 119(5): 874-892.

Chu, J. Lo, S-C. R. & Lee, I. K. 1993. Instability of granular soils under strain path testing. *Journal of Geotechnical Engineering* 119(5): 874-892. ASCE.

Chu, J. & Leong, W. K. 2001. Pre-failure strain softening and pre-failure instability of sand: a comparative study. *Geotechnique* 51(4): 311-321.

Chu, J. & Leong, W. K. In press. Effect of fines on instability behaviour of loose sand. *Geotechnique*

Drucker, D. C. & Seereeram, D. 1987. Remaining at yield during unloading and other unconventional elastic-plastic response. *Journal of Applied Mechanics* ASME. 54(1): 22-26.

Eckersley, J. D. 1990. Instrumented laboratory flowslides. *Geotechnique* 40(3): 489-502.

Fleming, R. W. Ellen, S. D. & Algus, M. A. 1989. Transformation of dilative and contractive landslide debris into debris flows – an example from Marin County, California. *Engineering Geology* 27: 201-223.

Gajo, A. Piffer, L. & De Polo, F. 2000. Analysis of certain factors affecting the unstable behaviour of saturated loose sand. *Mechanics of Cohesive and Frictional Materials* 5: 215-237.

Harp, E. W. Weels, W. G. II & Sarmiento, J. G. 1990. Pore pressure response during failure in soils. *Geol. Soc. Am. Bull.* 102(4): 428-438.

Jefferies, M. 1997. Plastic work and isotropic softening in unloading. *Geotechnique* 47(5): 1037-1042.

Lade, P. V. Nelson, R. B. & Ito, Y. M. 1988. Instability of granular materials with non-associated flow. *Journal of Engineering Mechanics* ASME. 114(12): 2173-2191.

Lade, P. V. & Pradel, D. 1990. Instability and plastic flow of soils. I: Experimental Observations. *Journal of Engineering Mechanics* ASME. 116(11): 2532-2550.

Lade, P. V. 1992. Static Instability and liquefaction of loose fine sandy slopes. *Journal of Geotechnical Engineering* ASME. 118(1): 51-71.

Lade, P. V. 1993. Initiation of static instability in the submarine Nerlerk berm. *Canadian Geotechnical Journal* 30(6): 895-904.

Leong, W. K. Chu, J. & Teh, C. I. 2000. Liquefaction and instability of a granular fill material. *Geotechnical Testing Journal* ASME. 23(2): 178-192.

Leong, W. K. 2001. Instability behaviour of a granular fill material. *Ph.D. Thesis* Nanyang Technological University, Singapore.

Leong, W. K. & Chu, J. 2002. Effect of undrained creep on instability behaviour of loose sand. *Canadian Geotechnical Journal* 39(6): 1399-1405.

National Research Council 1985. Liquefaction of soils during earthquakes. *Committee on Earthquake Engineering, Commission on Engineering and Technical Systems* Washington D C: National Academy Press.

Olson, S. M. Stark, T. D. Walton, W. H. & Castro, G. 2000. 1907 static liquefaction flow failure of the north dike of Wachusett dam. *Journal of Geotechnical and Geoenvironmental Engineering* ASCE. 126(12): 1184-1193.

Sasitharan, S. Robertson, P. K. Sego, D. C. & Morgenstern, N. R. 1993. Collapse behaviour of sand. *Canadian Geotechnical Journal* 30(4): 569-577.

Sladen, J. A. D'Hollander, R. D. Krahn, J. & Mitchell, D. E. 1985b. Back analysis of the Nerlerk berm liquefaction slides. *Canadian Geotechnical Journal* 22(4): 579-588.

Terzaghi, K. & Peck, R. B. 1967. *Soil mechanics in engineering practice*, 2nd (ed.), New York: John Wiley & Sons.

Wan, R.G. and Guo, P.J. 1998. A simple constitutive model for granular soils: modified stress-dilatancy approach. *Computers and Geotechnics* 22: 109-133.

Bifurcations & Instabilities in Geomechanics, – Labuz & Drescher (eds.)
© 2003 Swets & Zeitlinger, Lisse, ISBN 90 5809 563 0

Shear band displacements and void ratio evolution to critical state in dilative sands

A.L. Rechenmacher
The Johns Hopkins University

R.J. Finno
Northwestern University

ABSTRACT: Local displacements and void ratio evolution to critical state were evaluated for dilative sand specimens using the technique of Digital Image Correlation (DIC), which operates by mapping pixel movements between digital images. High-resolution digital images were taken of in-plane displacements throughout plane strain compression. The evolution of displacement from uniform to bifurcation to failure was quantitatively analyzed, providing insight regarding the point of onset of bifurcation. Displacement points within a shear band were isolated for strain computation, enabling assessment of critical states. The dependency of Critical State Line (CSL) position in void ratio-effective stress space on deposition void ratio was investigated by consolidating groups of specimens from similar initial void ratios. Uniformity of volumetric strain along the length of a shear band was investigated. Shear band thickness and inclination were measured. Results suggest a dependency of CSL position on initial or deposition void ratio.

1 INTRODUCTION

It is well known that shear bands form in dilative sands at or near peak stress and that the post-peak deformation regime is thus non-uniform. Difficulties in experimentally capturing quantitatively the details of the bifurcation process pose difficulty in elemental behavioral analyses, such as the evaluation of true constitutive behavior and quantification of the critical state. The uniqueness of the Critical State Line (CSL) in void ratio-effective stress space is a well-excepted concept. However, the formation of shear bands in dense sands at peak stress has in the past inhibited inclusion of such specimens in assessments of critical state. As such, the uniqueness of the CSL in void ratio-effective stress space has not been verified over the entire range of void ratios. Recent research, which has attempted to consider shear band behavior, has suggested a dependence of CSL position on initial or deposition void ratio.

Recent experimental advances have contributed to improved understanding of localization behavior in sands. X-ray Computed Tomography (CT) has shown promise in discerning shear band patterns in axisymmetric specimens (Desrues et al., 1996 and Alshibli et al., 2000). With this technique, however, the specimen must be placed in a separate scanner for the CT analysis, thus disrupting deformation and limiting confining pressure magnitude to vacuum pressure. Microscopic observation of thin slices through shear bands taken from epoxy-hardened specimens has recently been performed (e.g. Oda and Kazama, 1998 and Jang and Frost, 2000) and has availed thorough characterization of the microstructure within and in the vicinity of shear bands. However, only the hardened, failed state can be evaluated. Stereophotogrammetry has been applied to delineate shear band limits in photographs of in-plane behavior in plane strain tests (Finno et al., 1996; Finno et al., 1997; Mooney et al., 1997 and Mooney et al., 1998). Individual sand grain locations were digitized to track localized displacements. The technique was, however, tedious to perform.

More recently, the technique of Digital Image Correlation (DIC) has yielded highly accurate, *direct* quantification of the evolution of localized displacements in plane strain specimens throughout compression (Finno and Rechenmacher, 2003 and Rechenmacher and Finno, 2003). The DIC technique operates by matching pixels between consecutive digital images of a deformation process. By overlapping pixel mapping groups, large amounts of intensely spaced local displacement data are obtained.

This paper presents an overview of the results of DIC-derived local displacement measurements in dilative sand specimens undergoing plane strain compression. The evolution of specimen displacement from uniform to bifurcation to failure is quantitatively described and compared with globally measured stress-strain behavior, thus providing insight regarding the point of onset of strain localization. The displacement information derived from the DIC measurements was used to quantify volumetric evolutions to critical state in dilative sands, allowing accurate evaluation of CSL position in void ratio-effective stress space for a wide range of consolidated states. The derived CSL was not unique. Rather, more than one CSL was evidenced, suggesting a dependency of CSL position on initial or deposition void ratio. Additionally, the uniformity of deformation along the length of a persistent shear band was evaluated and will be discussed. Measurements of shear band thickness and inclination are presented as well.

2 EXPERIMENTAL TECHNIQUES

2.1 *Experimental program*

Experiments were conducted in a plane strain testing apparatus configured to initiate persistent shear banding (Vardoulakis and Drescher, 1988). The apparatus tests a prismatic-shaped specimen (140- by 40- by 80-mm) confined between two rigid, glass-lined and lubricated sidewalls. The glass-lined bottom platen rests in a low-friction, linear bearing sled, the configuration of which promotes unconstrained shear band propagation and growth. One of the sidewalls is constructed of 25-mm-thick, clear Plexiglas, enabling photographic observation of in-plane shear band behavior (Harris et al., 1995 and Finno et al., 1996). High-resolution digital images (1024 x 1024 pixels) were taken at various strain intervals (typically every 0.5 to 1.0 % axial strain) throughout deformation, using a Kodak Megaplus 4.2 experimental-grade digital camera.

Drained, plane strain, strain-controlled compression tests were performed on two different sands (Table 1). Both sands contained multi-colored grains that provided sufficient gray level contrast in the images to produce the uniquely patterned pixel subsets required by the DIC technique.

Table 1. Sand Properties

	Mason Sand	Concrete Sand
D_{50}	0.32 mm	0.62 mm
C_u	1.3	3.8
C_c	1.02	0.67
G_s	2.67	2.74

2.2 *Digital Image Correlation (DIC)*

The Digital Image Correlation (DIC) technique measures surface (2-D) displacements from digital images by matching pixel gray level values, or intensity values, between two digital images taken at different times in a deformation process (Sutton et al., 2000; Bruck et al., 1989). For an 8-bit image, gray levels range numerically between 0 (pure black) and 255 (pure white). Square subsets of pixels are matched, rather than individual pixels, as subsets comprising a wide variation in gray levels are more uniquely identifiable. By overlapping the pixel subsets, full-field displacement information is obtained.

194

DIC aims to find the displacements, u and v, of the center of each pixel subset. If the center of a subset displaces an amount u and v, then the initial coordinates, x and y, of any other pixel in that subset are related to their displaced coordinates in a later image, x^* and y^*, by:

$$x^* = x + u + \frac{\partial u}{\partial x}\Delta x + \frac{\partial u}{\partial y}\Delta y \tag{1}$$

$$y^* = y + v + \frac{\partial v}{\partial x}\Delta x + \frac{\partial v}{\partial y}\Delta y \tag{2}$$

where Δx and Δy are distances from the subset center to point (x,y). The inclusion of partial derivatives of the displacements allows for small straining of the subset.

A normalized cross-correlation coefficient, S, is used to measure the correspondence of pixel intensity values between two subsets (e.g., Bruck et al., 1989):

$$S\left(x, y, u, v, \frac{\partial u}{\partial x}, \frac{\partial u}{\partial y}, \frac{\partial v}{\partial x}, \frac{\partial v}{\partial y}\right) = 1 - \frac{\sum[F(x, y) \times G(x^*, y^*)]}{\left\{\sum[F(x, y)^2] \times \sum[G(x^*, y^*)^2]\right\}^{\frac{1}{2}}} \tag{3}$$

where $F(x,y)$ represents the gray level value at a coordinate in the initial image, and $G(x^*, y^*)$ the gray level value at a point in the deformed image. The values for u, v, $\partial u/\partial x$, $\partial u/\partial y$, $\partial v/\partial x$, and $\partial v/\partial y$ that minimize S will represent the best estimate of a subset's displacement. Typically, gray levels are interpolated between pixels, thus producing a continuous intensity function, thereby enabling displacement measurement to sub-pixel accuracy (Correlated Solutions, 2002). The program VIC-2D, by Correlated Solutions, Inc., was used in the current research to measure local displacements via the DIC technique. Accuracy tests within the current biaxial system (Rechenmacher and Finno, 2003) have indicated displacement measurement accuracy to within ±0.08 mm.

2.3 Local Displacement Measurement

Local displacements were measured incrementally between consecutive image pairs. DIC-based local displacement analyses were not possible over the entire specimen face due to minor membrane whitening around the specimen edges and some hardware interference. Thus, typically, an area encompassing about 2/3 of the specimen was analyzed. When shear band volumetric strain computations were of interest within a persistent shear band, only the region surrounding the shear band was analyzed to limit the amount of data.

As will be shown below, by plotting contours of the measured displacements, the location of a persistent shear band was easily identified. To determine shear band volumetric strains, and thus specimen evolution to Critical State, displacement points falling within the identified the shear band limits were isolated and then fit to a statistical-based displacement function (Mooney et al., 1997). Local strains then were computed from the derivatives of this function using a large-strain formulation.

3 LOCAL DISPLACEMENT EVOLUTION

Figure 1 shows typical global stress and local lateral strain behavior measured during a biaxial test on a mason sand specimen, but the results are typical of those observed for both sands. Lateral strain was measured from two pairs of displacement transducers affixed to the specimen as illustrated. The data points in Figure 1a indicate typical image locations.

Figure 2 shows incremental displacement vectors and contours obtained from DIC analyses performed consecutively between the photographs indicated in Figure 1a. For convenience, measured displacements were referenced to the bottom of the specimen by subtracting the magnitude of sled movement as measured by a corresponding LVDT. The scale bar represents distance along the specimen both in horizontal and vertical directions. The DIC analyses for this

195

mason sand test each resulted in about 3100 displacement points across the analysis area. For clarity, only about 5 to 10 percent of the displacement vectors are shown.

In the first three photographic increments, deformation appears to be roughly homogeneous, uniform, and symmetric, as indicated by the nearly horizontal and parallel contour lines. The fourth photographic increment, from 3.03 to 3.84% axial strain, encompassed just barely the achievement of peak stress (peak occurred at 3.67% axial strain, indicated by the vertical dashed line in Figure 1a). The local displacement contour for this increment no longer reveals uniform displacement as in the previous increments, but rather a slight tendency of back-to-back "V"-shaped contouring, with a slightly more intense displacement gradient oriented downward to the left. This pattern perhaps indicates that near peak stress deformations tend to focus along conjugate planes.

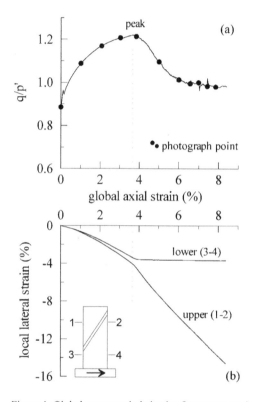

Figure 1. Global stress-strain behavior for mason sand specimen M14.

By the next two axial strain increments (3.84 to 5.0 % and 5.0 to 6.0%), however, deformation has almost entirely concentrated along a persistent shear band, oriented downward to the left. Remnants of concentrated displacement oriented in an approximately conjugate direction (downward to the right) can still be seen. Since the majority of specimen movement has concentrated within the shear band, shear band behavior thus must be considered in any characterizations of specimen behavior beyond peak stress.

The local lateral strain data in Figure 1b indicates that the shear band intercepted the upper LVDT pair. As indicated in Figure 1b, the curvature of the line measuring lower lateral strain reverses at a point which immediately precedes the achievement of peak stress. Finno et al. (1997) and Mooney et al. (1998) have associated this point of curvature reversal with the onset of shear formation. The corresponding displacement contour (3.03-3.84%) derived from DIC confirms that nonuniform deformation has indeed commenced within the strain increment that captured peak stress. A closer photographic spacing may have revealed the point of onset more precisely.

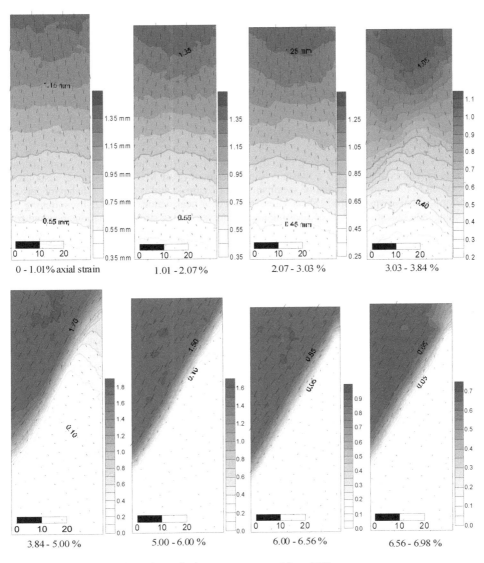

Figure 2. Local incremental specimen displacements measured from DIC.

A comparison between the DIC contours and the local lateral strain data from the upper LVDT pair provides insight regarding when the shear band achieves full formation. From Figure 1b, the rate of width expansion of the upper LVDT pair appears to reach a constant slope just after about 6.0% axial strain. Not coincidently, the local deformation contours indicate full concentration of specimen deformation within the shear band in the DIC analysis increments after 6.0% axial strain (note that there were no displacement data points in the upper right corner of the specimen in the last two increments, thus leading to slightly erratic contouring in that region [an artifact of the contouring program]).

The contours after peak stress indicate that almost all of the significant specimen deformation (the "pockets" of nonuniform displacment remote from the shear band are of the same magni-

197

tude as displacement measurement accuracy) has concentrated within a narrow zone of linear displacement. This zone defines the shear band. These contours provided a unique format for delineating the shear band limits. The full field of displacement data, produced by the overlapping nature of the DIC analysis subsets, reveals that the shear band limits are indeed very well defined. By plotting the displacement contours to scale, one can fairly objectively estimate thickness and inclination. Details of the shear band inclination and thickness measurements will be discussed below.

4 STRESS-STRAIN BEHAVIOR AND CRITICAL STATES

4.1 *Global and Local Responses*

Figure 3 shows the shear stress, q, effective stress ratio, q/p', volumetric strains and dilatancy versus global axial strain for a masonry sand specimen (Test M17). The global response of specimen M17 is similar to that of specimen M14 (Figure 1) and is typical of behavior for very uniform, dilative specimens. Shear stress achieves a peak value, followed by softening and eventual attainment of a state of shearing at constant stress (the stress fluctuations resulted from shearing rate changes imposed as part of a companion study). Volumetrically, the specimen initially contracts but then dilates. Eventually the specimen appears to have achieved a critical state of continued shearing at constant stress and volume. However, as indicated in Figures 1b and Figure 2, once peak stress is attained and a persistent shear band has formed, global measurements no longer reflect accurately true specimen behavior. Local behavior needs to be considered to properly characterize specimen behavior.

The results of the shear band volumetric strain computations using DIC-measured displacement points are shown in Figure 3b and are compared with the volumetric strains obtained from conventional global measurements. Prior to peak stress, when the specimen was globally deforming uniformly, approximately 1.0% volumetric dilation had occurred. Following peak stress, volumetric strains were computed using only the DIC-derived displacement points within the shear band, the only part of the specimen involved in significant deformation. These local computations indicate that almost 8.0% volumetric dilation had occurred upon the achievement of critical state. This contrasts markedly the roughly 2.0% of volumetric dilation that was recorded globally. This disparity of measurements highlights the need to consider the volumetric evolution with the shear band to enable precise quantification of critical state for dense, dilative specimens. Figure 3c shows that despite the difference in global versus local volume change, the dilatancy angle reaches zero and the soil within the shear band indeed reaches a critical state.

4.2 *Critical States*

To assess the dependency of Critical State Line (CSL) position on deposition void ratio, specimens of each sand were consolidated in two distinct groups. Three to four specimens were consolidated from a similar initial medium dense void ratio, each to a different anisotropic stress level. Then, three to four specimens were consolidated from similar dense initial void ratios. In this way, two separate, distinct consolidation histories would be formed, each representing consolidation from a distinct initial or "deposition" void ratio.

The consolidated states of all specimens are shown by the open symbols in Figure 4. For the mason sand specimens in Figure 4a, three medium-dense specimens were all consolidated from initial void ratios between 0.75 and 0.76. Four dense specimens each were consolidated from initial void ratios ranging between 0.66 and 0.68. Thus, two distinct initial, or deposition void ratios are represented. Similarly, for the concrete sand specimens, three specimens were consolidated from medium dense initial void ratios between 0.59 and 0.60, and three specimens were consolidated from dense initial void ratios ranging between 0.53 and 0.54.

Stresses and void ratios at the critical states were determined from a synthesis of pre-peak global and post-peak local responses. The resultant critical states and inferred CSLs in void ratio-effective stress space are shown in Figure 4. The same symbol shape is used both for consolidated and critical states. For example, the consolidated state for specimen M14 is indicated by the open square.

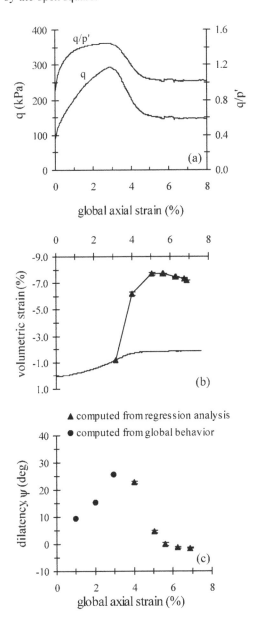

Figure 3. Typical global and local stress and volumetric responses (specimen M17).

The CSLs for the two deposition void ratios of the mason sand are shown in Figure 4a, along with the steady state line derived from undrained, plane strain compression tests performed on loose samples of the same sand (Finno et al., 1996). The three lines are separate: the CSL formed by the lower-void ratio samples of the denser consolidation history plots below, or to the left of, the CSL for the looser samples of the medium-dense consolidation history. Additionally, the two CSLs and the undrained steady state line are approximately parallel, in accordance with Poulos, et al. (1985), who indicated that the slope of the CSL depends on the shape of the grains of a given soil. Because all specimens, regardless of deposition void ratio, were derived from the same soil, and thus the same grain shape, parallel CSLs are expected.

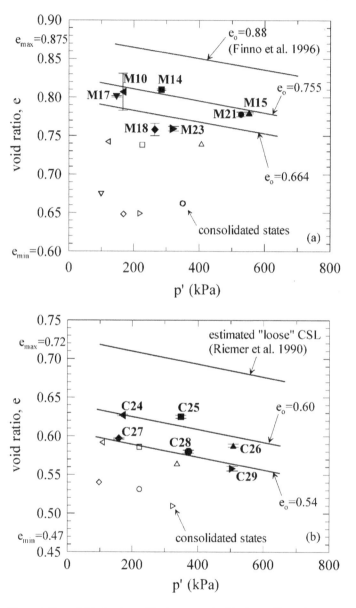

Figure 4. Consolidated states, critical states and critical state lines: a) mason sand and b) concrete sand.

The CSLs for the two different deposition void ratios of the concrete sand are shown in Figure 4b. Also shown is an estimated CSL for loose specimens, based on the suggested procedure of Riemer et al. (1990), who observed that the maximum void ratio forms an upper limit for the CSL. The estimated line was drawn parallel to the other two lines starting at the maximum void ratio and an effective stress of 10 kPa. While the actual location of this estimated CSL for loose specimens is unknown, any reasonable estimate using the maximum void ratio would clearly place it above the two lines found for the denser concrete sands. As with the masonry sands, the two experimental lines are separate and parallel, with the denser initial states having a lower CSL than the medium dense states.

The separation of the CSLs on Figure 4 suggests that CSL position is a function of deposition void ratio, or consolidation history, and hence depends on the initial state of the sand. These results emphasize the importance of obtaining high quality samples of sands, and thus replicating the *in situ* structure of a soil as much as possible, when conducting experiments to determine critical state of a natural soil deposit.

5 SHEAR BAND UNIFORMITY

A typical DIC analysis produced between 100 and 600 displacement points within a shear band. The large number of displacement points produced by the DIC technique permitted analysis of the uniformity of volumetric strain behavior along the length of a persistent shear band. Previously, Mokni (1992) found erratic volumetric behavior along the length of shear bands. Finno et al. (1997) and Mooney et al. (1997) found that variations in volumetric strains occurred locally within a shear band, but that the shear band was deforming uniformly in an average sense.

Shear band uniformity analyses are presented for a mason sand specimen (Figure 5) and a concrete sand specimen (Figure 6). DICs were performed on the middle 70% of the shear band for test M23, and the middle 65% for test C27 (membrane discoloration near the edges of the specimen prevented analysis of the full length). In both cases, the persistent shear band formed downward to the right. DIC analyses for test M23 typically resulted in about 300 displacement points across the shear band. These points were divided into three separate analysis regions: upper left 1/3, middle 1/3, and lower right 1/3. Care was taken to include at least 80 displacement points in each analysis region, as Finno and Rechenmacher (2003) determined that at least 80 points were necessary to converge to accurate volumetric strain quantification. DIC analyses for test C27 produced about 200 displacement points across the shear band. To keep the number of displacement points per analysis region above 80, the shear band was divided into three overlapping regions: upper left 1/2, middle 1/2, and lower right 1/2.

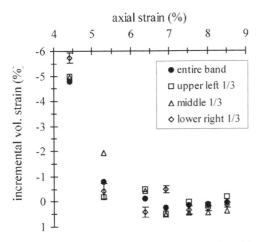

Figure 5. Shear band incremental volumetric strain uniformity during test M23.

Incremental volumetric strains for the three different regions across the shear band and for the entire band using all displacement points are shown in Figures 5 and 6 for tests M23 and C27, respectively, for each photographic increment following peak stress. For clarity, error bars for volumetric strains are shown for one region only. In initial strain increments after peak stress but prior to the achievement of critical state (approximately where zero incremental volumetric strain, on average, is first attained), volumetric behavior across the band was generally erratic, with about 1 to 2% variation in computed strains among the different regions. This variation is significant, considering that volumetric dilation at critical state was on average about 4%. The magnitude and direction of this variation was not consistent: no one region showed consistently more or less volumetric activity. Once critical state was achieved, the volumetric variations among the regions became less pronounced, differing in most cases by less than 0.5%, which is on the order of the error in strain computation.

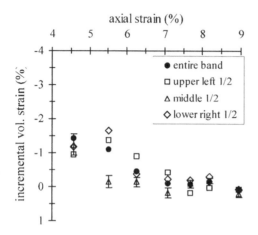

Figure 6. Shear band incremental volumetric strain uniformity during test C27.

The results in Figures 5 and 6 show that strain computations based on analysis points spread over the "entire band" produce an "average" value among the different regions, which verifies the results of Finno et al. (1997) and Mooney et al. (1997). Therefore, when strain computations (via the regression technique) are based on at least 80 displacement points that are derived from the entire shear band length, consistent results can be obtained.

6 SHEAR BAND THICKNESS AND INCLINATION

Measurements were recorded at peak stress of shear band inclination, β, referenced relative to the direction of the minor principal stress, σ_3, which in the case of biaxial compression was horizontal. For comparison, the theoretical solutions from Mohr-Coulomb theory, β_C, and from Roscoe (1970), β_R, were computed, as was the empirically-based expression derived by Arthur et al. (1977), β_A:

$$\beta_C = 45 + \frac{\phi'}{2} \tag{4}$$

$$\beta_R = 45 + \frac{\psi}{2} \tag{5}$$

$$\beta_A = 45 + \frac{(\psi + \phi')}{2} \tag{6}$$

The values ϕ' and ψ were taken at peak stress so that the comparison among the predictions and measured data would be consistent.

Measured and predicted shear band inclinations at first formation (taken as coincident with peak stress) are compared in Figure 7. Measured orientations for the mason sand range between 59° and 64°, in agreement with previously reported values of 55° to 65° for drained and undrained tests on the same sand (Finno et al., 1997). However, values reported here are more narrowly bounded, which is perhaps a consequence of the improved measurement accuracy. Concrete sand inclinations ranged between 53 and 59°. For both sands, shear band orientations were best predicted by the Arthur solution.

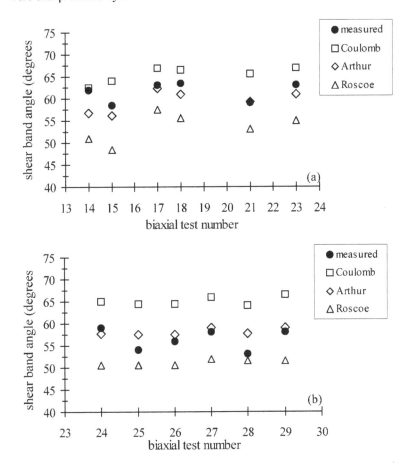

Figure 7. Measured and predicted shear band inclinations at peak stress: a) mason sand and b) concrete sand.

The observed trend of lower shear band inclination for larger median grain size agrees with the work of others. In accordance with Arthur and Dunston (1982), who found that β decreased from $β_C$ to $β_R$ as particle size increased, the mason sand shear band inclinations tended to be bounded by the Coulomb and Arthur solutions, while inclinations in larger-grain-size concrete sand samples tended to be bound by Arthur and Roscoe solutions. The results for the concrete sand additionally agree with the work of Han and Drescher (1993), who measured inclinations falling between the Arthur and Roscoe solutions for a sand of $D_{50} = 0.72$ mm, which was similar to the mean grain size of the concrete sand. Mokni (1992) and Alshibli and Sture (2000) also measured lower shear band inclinations for sands with larger grain sizes.

Measured shear band inclinations at peak and critical state for both mason and concrete sand samples are compared in Figure 8. In all cases, the shear band became more horizontal as shearing progressed, on average decreasing by about 2° from peak to critical state.

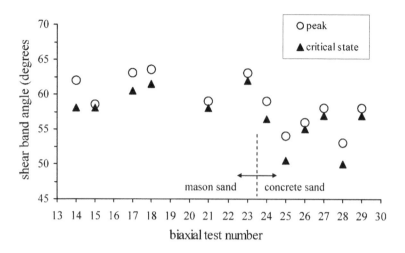

Figure 8. Measured shear band inclinations at peak and critical state.

Figure 9 summarizes for all tests the shear band thickness, t, normalized by the mean grain diameter, D_{50}, measured at peak stress and first achievement of critical state. For the mason sand specimens, average shear band thickness was 17 times D_{50} at peak and 18 times D_{50} at critical state. The average thicknesses for concrete sand specimens were 15 times D_{50} at peak and 17 times D_{50} at critical state. These magnitudes are consistent with those of Finno et al. (1997), who reported thicknesses between 10 and 25 times D_{50} for the same mason sand. The range of thicknesses measured herein is, again, generally narrower. The results also agree with Liang et al. (1997), who measured for a uniform fine sand ($D_{50} = 0.26$ mm) average thicknesses between 13 and 21 times D_{50}.

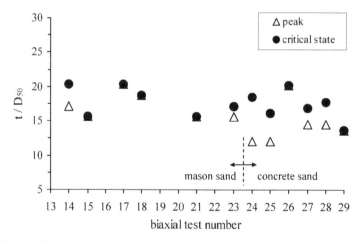

Figure 9. Normalized shear band thickness measurements at peak stress and critical state.

The data in Figure 9 indicate that growth in shear band thickness during a test was more significant for concrete sand specimens than mason sand specimens. For the mason sand specimens, shear band formation occurred much more abruptly, and in most cases, most of the measurable shear band growth occurred within the photographic increment encompassing peak stress. In contrast, shear bands in concrete sand specimens formed more slowly, thus accounting for the notable thickness changes observed between peak and critical state.

7 CONCLUSIONS

Digital Image Correlation (DIC) has been shown to be a detailed and precise method for discerning local displacements in sands undergoing plane strain compression. Large amounts of displacement data are obtained and have been used to evaluate patterns of shear band formation, local shear band volumetric evolutions to critical state, local deformation uniformity, and shear band thickness and inclination measurements.

The following observations have been made.
1. The onset of shear band formation appears to precede the achievement of peak stress. Onset is characterized by a tendency toward conjugate band formation, but a single shear band quickly dominates.
2. Critical states measured from dilative specimens suggest that the position of the Critical State Line in void ratio-effective stress space is not unique for a given sand, but rather is dependent upon deposition void ratio
3. In strain increments after peak stress, but prior to critical state, volumetric deformation within a persistent shear band was quite erratic. Behavior became fairly uniform once critical state was achieved.
4. Shear band inclinations were best predicted by the Arthur solution.

8 REFERENCES

Alshibli, K.A. & Sture, S. 2000. Shear Band Formation in Plane Strain Compression. *Journal of Geotechnical and Geoenvironmental Engineering* 126 (6): 495-503.
Alshibli, K.A., Sture, S., Costes, N.C., Frank, M.L., Lankton, F.R., Batiste, S.N. & Swanson, R.A. 2000. Assessment of Localized Deformations in Sand Using X-Ray Computed Tomography. *Geotechnical Testing Journal* 23 (3): 274-299.
Arthur, J.R.F. & Dunston, T. 1982. Rupture Layers in Granular Media. *Proceedings IUTAM Conference on Deformation and Failure of Granular Materials, Delft.* Rotterdam: Balkema. 453-459.
Arthur, J.R.F., Dunstan, T., Al-Ani, Q.A.J.L. & Assadi, A. 1977. Plastic Deformation and Failure in Granular Media. *Geotechnique* 27 (1): 53-74.
Bruck, H.A., McNeill, S.R., Sutton, M.A. & Peters III, W.H. 1989. Digital Image Correlation Using Newton-Raphson Method of Partial Differential Corrections. *Experimental Mechanics* 29 (3): 261-268.
Correlated Solutions 2002. Correlated Solutions Web Page http://www.correlatedsolutions.com.
Desrues, J., Chambon, R., Mokni, M. & Mazerolle, F. 1996. Void Ratio Evolution Inside Shear Bands in Triaxial Sand Specimens Studied by Computed Tomography. *Geotechnique* 46 (3): 529-546.
Finno, R.J., Harris, W.W., Mooney, M.A. & Viggiani, G. 1996. Strain Localization and Undrained Steady State of Sand. *Journal of Geotechnical Engineering* 122 (6): 462-473
Finno, R.J., Harris, W.W., Mooney, M.A. & Viggiani, G. 1997. Shear Bands in Plane Strain Compression of Loose Sand. *Geotechnique* 47 (1): 149-165.
Finno, R.J. & Rechenmacher, A.L. 2003. The Effect of Consolidation History on the Critical State of Two Sands. *Journal of Geotechnical and Geoenvironmental Engineering* in print
Han, C. & Drescher, A. 1993. Shear Bands in Biaxial Tests on Dry Coarse Sand. *Soils and Foundations* 33 (1): 118-132.
Harris, W.W., Viggiani, G., Mooney, M.A. & Finno, R.J. 1995. Use of Stereophotogrammetry to Analyze the Development of Shear Bands in Sand. *Geotechnical Testing Journal* 18 (4): 405-420.
Jang, D.-J. & Frost, J.D. 2000. Use of Image Analysis to Study the Microstructure of a Failed Sand Specimen. *Canadian Geotechnical Journal* 37: 1141-1149

Liang, L., Saada, A., Figueroa, J.L. & Cope, C.T. 1997. The Use of Digital Image Processing in Monitoring Shear Band Development. *Geotechnical Testing Journal* 20 (3): 324-339.

Mokni, M. 1992 *Relations Entre Deformations en Masse et Deformations Localisees dans les Materiaux Granulaires*. These de Doctorat de l'Universite J. Fourier de Grenoble, France.

Mooney, M.A., Viggiani, G. & Finno, R.J. 1997. Undrained Shear Band Deformation in Granular Media. *Journal of Engineering Mechanics* 123 (6): 577-585.

Mooney, M.A., Finno, R.J. & Viggiani, G. 1998. A Unique Critical State for a Sand? *Journal of Geotechnical and Geoenvironmental Engineering* 124 (11): 1100-1108.

Oda, M. & Kazama, H. 1998. Microstructure of Shear Bands and its Relation to the Mechanisms of Dilatancy and Failure of Dense Granular Soils. *Geotechnique* 48 (4): 465-481.

Poulos, S.J., Castro, G. & France, J.W. 1985. Liquefaction Evaluation Procedure. *Journal of Geotechnical Engineering* 111 (6): 772-791.

Rechenmacher, A.L. & Finno, R.J. 2003. Digital Image Correlation to Evaluate Shear Banding in Dilative Sands. *Geotechnical Testing Journal* in print.

Roscoe, K.H. 1970. The Influence of Strains in Soil Mechanics. *Geotechnique* 20 (2): 129-170.

Sutton, M.A., McNeill, S.R., Helm, J.D. & Chao, Y.J. 2000. Advances in Two-Dimensional and Three-Dimensional Computer Vision. *Photomechanics, Topics in Applied Physics* 77: 323-372.

Vardoulakis, I. & Drescher, A. 1988. Development of Biaxial Apparatus for Testing Frictional and Cohesive Granular Media. *Report to National Science Foundation, NSF Grant No. CEE84-06500*. University of Minnesota, Duluth, MN.

Bifurcations & Instabilities in Geomechanics, – Labuz & Drescher (eds.)
© *2003 Swets & Zeitlinger, Lisse, ISBN 90 5809 563 0*

Plane strain compression of sandstone under low confinement

J.J. Riedel & J.F. Labuz
Department of Civil Engineering, University of Minnesota, Minneapolis, USA

Shong-Tao Dai
Minnesota Department of Transportation, St. Paul, Minnesota, USA

ABSTRACT: Closed-loop, servo-controlled experiments were conducted to investigate the evolution of the shear band in a sandstone under low confinement. The tests were performed with a plane-strain apparatus designed to allow the failure plane to develop and propagate in an unrestricted manner. Thin-section microscopy provided direct observations in and adjacent to the shear band. Two distinct regions of the failure surface were recognized from low-confinement experiments. The first region, called the primary fracture, cut diagonally through the center of the specimen and stopped before intersecting the surface. The second region featured a kink and began nearly at the termination of both ends of the primary fracture. The kinked region was inclined at a less steep angle to the direction of minimum compressive stress. Microscopic observations showed that the primary fracture was formed in shear and the secondary kinks were formed in tension.

1 INTRODUCTION

From laboratory compression tests on geomaterials such as overconsolidated clays, dense sands, and rocks, it is observed that a smoothly varying deformation pattern changes and that further deformation is localized in narrow regions called shear bands. This phenomenon is also observed in field situations, for instance, during collapse of slopes and rupture of the earth. Although the formation and orientation of the shear band may be described by equilibrium bifurcation theory (Vardoulakis & Sulem 1995), it has been demonstrated for rock that an interaction of tensile microcracks may lead to macroscopic shear failure (Dunn et al. 1973).

In sandstone, many authors have observed regions of altered porosity associated with localization of deformation and some have identified changes in porosity associated with shear band formation (Aydin 1978; Bernabe & Brace 1990; Ord et al. 1991; Menedez et al. 1996; Besuelle et al. 2000). In low confinement (pressure much less than the uniaxial strength) triaxial compression tests, dilatancy adjacent to the failure surface is a common feature, as are fractured grains within the localized zone. Some researchers have concluded that localization initiates in the hardening regime, while fracturing of grains and the formation of the failure mechanism usually occurs after peak stress in the softening regime.

A suite of plane-strain compression tests were performed on Berea sandstone with the University of Minnesota Plane Strain Apparatus (Labuz et al. 1996). Control of the test was maintained through peak and partially into the strain-softening regime. Visual and experimental observations corresponding to the localization of deformation in plane strain compression are reported. A through-going fracture surface evolved with two distinct regions characterizing failure. The first region, called the primary fracture, cut diagonally through the center of the specimen and stopped before intersecting the surface of the specimen. The second region began nearly at the terminations of both ends of the primary fracture, but was inclined or kinked at a less steep angle to the direction of minimum compressive stress. Microscopically, different modes of deformation within the two regions were observed: the primary fracture seemed to be formed in shear whereas the kinked regions seemed to be formed in tension.

2 EXPERIMENTS

A Vardoulakis-Goldscheider type plane-strain compression apparatus was designed and fabricated at the University of Minnesota. A sketch is shown in Figure 1, with a brief description of the various parts numbered [1-5] in the figure. The apparatus allows the failure plane within the specimen [1] to develop and propagate in an unrestricted manner by attaching the upper platen to a low friction linear bearing [2] referred to as the sled. Plane-strain deformation is enforced by a thick-walled steel cylinder called the biaxial frame [3]. A review of the plane-strain approximation involving a stiff biaxial frame follows from Labuz et al. (1996).

2.1 Biaxial frame

The interaction between the biaxial frame and the specimen is analyzed by approximating the frame as a structural element of stiffness k^f:

$$k^f = \frac{\sigma_{zz}^f A}{u_z^f},$$ (1)

where A is the cross-sectional area (in elevation view) of the specimen, σ_{zz}^f is the reaction stress from the frame, and $2u_z^f$ is the total displacement of the frame, which is equal to the total displacement of the specimen $2u_z$. The sign convention of compression positive is used:

$$\varepsilon_{zz} = -\frac{u_z}{w},$$ (2)

where $2w$ is the total width of the specimen. Within generalized Hooke's law, equilibrium between the specimen and the frame requires

$$\varepsilon_{zz} = \frac{1}{E}\left[-E^f \varepsilon_{zz} - v\left(\sigma_{xx} + \sigma_{yy}\right)\right],$$ (3)

where the modulus of the frame E^f is defined as

$$E^f = \frac{k^f w}{A}.$$ (4)

Note that the modulus of the frame is dependent on the area A of the specimen. Equation 3 can be rewritten to give

(a)   (b)

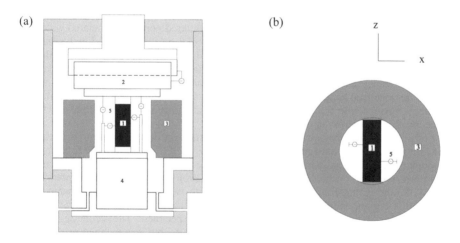

Figure 1. Sketch of the University of Minnesota Plane strain Apparatus. (a) Elevation. (b) Plan.

208

$$\varepsilon_{zz} = -\left(\frac{v}{E + E^f}\right)\left(\sigma_{xx} + \sigma_{yy}\right), \tag{5}$$

and from Equations 1, 4, and 5:

$$\sigma_{zz} = \frac{v\left(\sigma_{xx} + \sigma_{yy}\right)}{1 + E/E^f} \tag{6}$$

As expected, if $E^f \gg E$, then the plane-strain condition is realized.

To evaluate the effectiveness of the passive restraint in approximating the plane-strain condition, consider the ratio

$$R = -\frac{\varepsilon_{zz}}{\varepsilon_{yy}} \tag{7}$$

Upon substitution of Hooke's law and Equation 5 into 7 and simplifying, we get

$$R = \frac{v\left(\sigma_{xx} + \sigma_{yy}\right)}{\sigma_{yy} - v\sigma_{xx} + f\left(E^*\right)}, \tag{8}$$

where

$$f\left(E^*\right) = \frac{1+v}{E^*}\left[\sigma_{yy}(1-v) - v\sigma_{xx}\right] \text{ and } E^* = \frac{E}{E^f} \tag{9}$$

If the biaxial frame is soft, $E^* \to \infty$ and

$$R^{soft} = \frac{v\left(\sigma_{xx} + \sigma_{yy}\right)}{\sigma_{yy} - v\sigma_{xx}}. \tag{10}$$

Thus, for a passive restraint system, how well the conditions approximate plane strain can be expressed by percent plane strain (PPS):

$$PPS = \left(1 - \frac{R}{R_{soft}}\right) \times 100\%. \tag{11}$$

PPS of 100% is the case of perfectly rigid restraint.

The biaxial frame for the University of Minnesota Plane-Strain Apparatus (Fig. 1b) was designed as a thick-walled cylinder (OD = 300 mm, ID = 125 mm) of mild steel. A circular opening with flats, which provides precise alignment of the specimen, was machined in the frame. A nominal specimen width of 100 mm was chosen. Hardened steel platens of 6 mm thickness contacted the material and established kinematic (frictionless when lubricated) boundaries; the platens were enlarged, providing a surface to seal the specimen from the confining fluid. From tests with an aluminum specimen, the stiffness of the frame was estimated to be about 7 MN/mm.

2.2 Specimen setup

Prismatic specimens [1] are used, with the size range from 75-100 mm in height, 27-40 mm in thickness, and 100 mm in width. The axial load is measured by load cells located at the bottom [4] of the specimen and outside the pressure chamber. The axial displacement, the lateral displacement of both surfaces of the specimen exposed to confining pressure, and the lateral displacement of the top platen in contact with the sled are measured by LVDTs [5]. The two surfaces of the specimen exposed to confining pressure are sealed by a polyurethane coating; metal targets glued to the specimen provide firm contact points for the lateral LVDTs and acoustic emission sensors. The four surfaces in contact with polished-steel platens are covered with stearic acid to reduce the friction between the platens and the specimen (Labuz & Bridell 1993). Closed-loop, servo-controlled tests were performed on specimens of Berea sandstone with the average lateral displacement used as the feedback signal; the rate of loading was 10^{-4} mm/s and the confining pressure was 5 MPa.

The Berea sandstone tested was characterized by the following properties: $q_u = 40\text{-}45$ MPa, E = 15-16 GPa, $v = 0.32\text{-}0.34$ and an average grain size of 0.2 mm. Along with the measurements of deformation and acoustic emission, one specimen tested was analyzed using thin-section microscopy. The specimen, 99.30 mm in height and 27.84 mm in thickness, contained an imperfection, a 3 mm diameter hole drilled parallel to the plane-strain direction. The imperfection was used to cause the failure process to initiate at the center of the specimen, away from the corners.

To secure the specimen within the fixed inside diameter of the biaxial frame, a pair of wedges and corresponding spacer were designed. Slight variation (±0.15 mm) among specimens was allowed by matching a set of wedges and a spacer of the same thickness as the wedge assembly. The common wedge was fastened to one flat of the biaxial frame and the corresponding spacer was placed on the other flat. The specimen was secured into position by tapping the matching wedge to achieve a certain level of prestress (5-10 MPa); this was critical in eliminating any gaps between the specimen and the platens, so that the intermediate stress could develop fully. Four transducer-quality strain gages were glued with epoxy to the inside walls at the third points in height, placed so that their tangents were parallel to the z-axis. The biaxial frame was calibrated to give an estimate of the plane-strain stress based on the readings of the strain gages.

3 RESULTS

The axial force and lateral displacements versus the axial displacement response from the test is shown in Figure 2. After peak load, the specimen exhibited stable softening for a short period (an increment of axial displacement < 0.01 mm) followed by a class II snap-back response. It was possible to observe this portion of the load history because the average lateral displacement was the feedback signal. Nevertheless, an uncontrolled failure occurred before residual sliding and even before sled movement. The failure surface (Figure 3a) consisted of a primary feature with an average orientation of 77° and a less steep kinked feature at an average inclination of 62°. The resultant displacement vector calculated from the displacement measured by the lateral and axial LVDTs during residual sliding was oriented 70° from the minor principal stress direction. Thus, for the kinked fracture, 8° of closing occurred during the residual sliding portion of the test (Figure 3b), while the primary fracture experienced 7° of opening (Figure 3c).

Deformation can be conveniently described by volumetric strain ε_v and shear strain γ_s:

$$\varepsilon_v = \varepsilon_1 + \varepsilon_3 \qquad \gamma_s = \varepsilon_1 - \varepsilon_3 \tag{12}$$

which can be decomposed into increments of elastic and plastic components:

$$\Delta\gamma_s = \Delta\gamma^p + \Delta\gamma^e \qquad \Delta\varepsilon_v = \Delta\varepsilon^p + \Delta\varepsilon^e \tag{13}$$

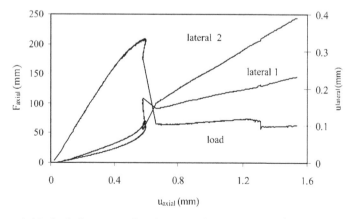

Figure 2. Mechanical response of sandstone specimen at 5 MPa confinement.

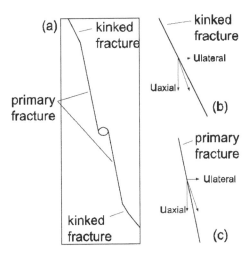

Figure 3. Sketch of the failure zone and kinematics of post-peak displacements.

The principle strain directions correspond to the Cartesian coordinates defined for the plane strain device, that is, $\varepsilon_{yy} = \varepsilon_1$ and $\varepsilon_{xx} = \varepsilon_3$.

The behavior of the material at low deviator stress (Figures 4a&b) showed an initial nonlinear response, which was attributed to seating effects. Afterwards, the material response was dominated by elastic behavior up to a stress difference of 52 MPa. The onset of plastic deformation was characterized by the nonlinear stress-strain response (Figures 4a&b) and it was associated with an increase of plastic volumetric strain (Figure 4c). The rate of dilation was approximately constant, with a dilation angle ψ, where $\sin \psi = -\Delta\varepsilon^p/\Delta\gamma^p$ of about 30°.

The sled did not move prior to peak load and even after peak (Figure 4d), meaning that the failure mechanism was not completely formed well into the post-peak regime. Furthermore, the volumetric strain continued to increase after peak stress and prior to the movement of the sled (Figure 4c). However, the incremental volumetric strain slightly decreased prior to the loss of control of the test (Figure 4c). The decrease may be a structural response rather than an element response, representing the combination of two competing deformation mechanisms, shear banding and elastic unloading.

3.1 Acoustic activity

The onset and development of the failure phenomenon was tracked with acoustic emission. The microseismic activity was monitored throughout the load history with eight piezoceramic sensors of 3 mm diameter. The acoustic emission (AE) signals were amplified (40 dB gain) and filtered (bandpass 0.1-1.2 MHz) before recording. The P-wave velocity was determined before the experiment. With four degrees of redundancy, the coordinates of the AE sources were obtained from the arrival times of the P-wave at the sensors by minimizing the sum of the squares of the distance residuals. A collapsing routine was used to accentuate the location of the failure plane.

The results show that the microseismic events emitted from 98-100% peak load appeared to show some localization (Figure 5a), with the AE clustered around the imperfection (the 3-mm diameter hole). In the post-peak regime from 100-90% of peak load, the AE activity extended farther away from the imperfection, and the shear band was propagating (Figure 5b). The acoustic emissions clearly indicated localization of deformation, although the events were not exactly positioned along the eventual rupture zone. From 90-80% in the post-peak region, the shear band continued to grow (Figure 5c) but the failure plane was not yet fully developed. Test control was then momentarily lost and an audible sound of rock fracture was heard. Thus, the steep portion of the failure plane formed first followed by the less steep kinked parts.

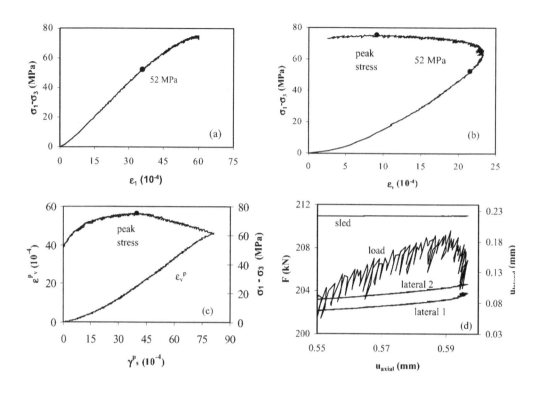

Figure 4. Mechanical behavior of Berea sandstone at 5 MPa confining pressure.

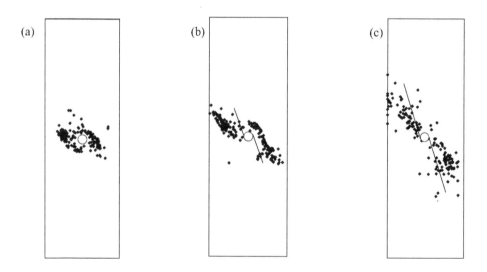

Figure 5. Acoustic emission locations. (a) Peak load. (b) About 90% post-peak. (c) About 80% post-peak.

3.2 *Optical microscopy*

Upon test completion, the specimen was coated with an epoxy glue and later potted in hydro-stone. After cutting the specimen in half, the block was quenched in a low viscosity blue-colored epoxy to increase the visibility of porosity and to preserve the structure of the failure plane during thin section preparation. Four thin-sections were prepared at equal intervals per-pendicular to the axis of plane strain. The thin-sections were examined under reflected light at magnifications ranging from about 10x – 50x. Photos were taken with a digital camera mounted on the microscope.

There was a visual increase in porosity along the margins of the primary feature extending about 2 mm or 10-grain diameters to each side (Figure 6a). The increase in porosity was inter-preted to have evolved in the hardening stage where grains translate and rotate pass each other. Grain motion and increase in porosity corresponded to an increase in plastic volumetric strain. In addition, micro-fractures were observed in individual grains along the inner most surface of the margin. Between the two margins, there was an observable decrease in particle size that was constituted by fractured or crushed grains (Figure 6a). Thickness of the crushed grains in the primary fracture varied from 0.2-0.5 mm.

 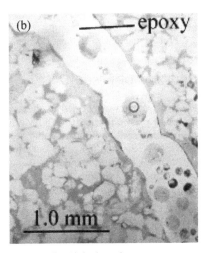

Figure 6. Photomicrograph of the failure zone. (a) Primary feature. (b) Kinked portion.

A kinked region, less inclined to the direction of minimum compression, illustrated a differ-ent mode of failure, not indicative of shearing: there was very little evidence of fracturing and crushing of grains along the margins of the failure surface (Figure 6b). The minor amounts of grain fracturing observed were attributed to sliding along the kinked surface after the failure mechanism was formed. In addition, there was no visual evidence for an alteration in porosity within the margins of the kinked region. The lack of crushed and fractured grains and the ab-sence of porosity alteration in the margins of the second region imply that the surface was formed by tensile fracture.

The change in inclination of the failure surface marked the transition between the two re-gions. Figure 7a illustrates the transition; the vertical line is parallel to the direction of maxi-mum compression. The top half of the kinked line indicates the general trend of the failure sur-face away from the transition zone, where the bottom half indicates the new direction. The quantity of the crushed and fractured grains in the image decrease as the failure surface is traced in the downward direction. Within the transition zone, the termination or process zone of the primary feature can be observed, with fragments of crushed material and fractured grains that continue off the primary fracture in a parallel direction and terminate. The termination of the primary fracture was most obvious in the thin-sections containing the largest change in inclina-tion at the kink (Figure 7b)

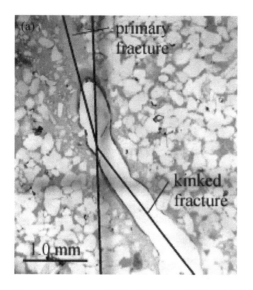 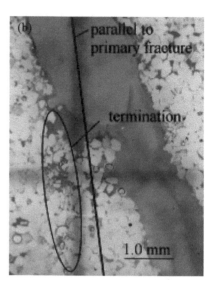

Figure 7. Primary and kinked features. (a) Transition region. (b) Termination zone.

4 DISCUSSION OF RESULTS

The primary feature was formed in shear, so the orientation of the failure plane can be investigated through plasticity theory. The Mohr failure envelope of the Berea sandstone was determined through a series of biaxial [b] and conventional triaxial [t] tests, with confining pressures of 0[t], 5[b], 7.5[b], 10[b], 20[t], and 40[t] MPa. As shown in Figure 8, a (parabolic) nonlinear yield surface of the form $\tau^2 = a + b\sigma$ fitted the data reasonably well. Enforcing the linear Coulomb criterion for the low-pressure experiments produced a friction angle $\phi = 54°$. Thus, the maximum orientation θ_c of the failure plane with respect to the minimum principal stress would give

$$\theta_c = 45° + \phi/2 \tag{14}$$

or 72°, less than the observed 77°. Analyses using hardening-softening plasticity models and equilibrium bifurcation (Arthur et al. 1977; Vardoulakis 1980) give the angle θ_b as

$$\theta_b = 45° + (\phi+\psi)/4 \tag{15}$$

which is always less than θ_c. Although a lubricant was used to reduce end-platen restraint, a small amount of friction can influence the inception of the shear band. The kinked portion of the failure surface may be a result of boundary effects (Needleman & Ortiz 1991), although the deformation mechanisms were identified as tensile in nature.

Linear fracture mechanics may offer another approach in analyzing the failure process. The controversial issue in this regard is the assumption of shear fracture, with initiation and propagation within the original crack plane. Nevertheless, two deformation states with known loading can be evaluated for a mode II stress intensity factor if a crack length can be assigned. In this respect, AE locations provided an estimate (Figures 5b,c) and a finite element analysis (FRANC2D 1995) provided K_{II}.

A general form of the stress intensity factor for the biaxial specimen is

$$\frac{K_{II}}{\sigma_3\sqrt{\pi a}} = Y(a/w)(k-1) \tag{16}$$

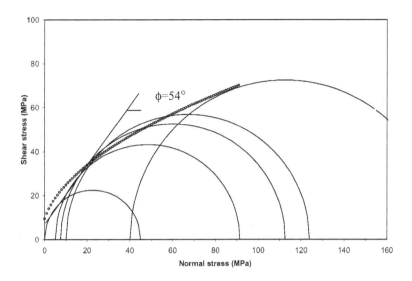

Figure 8. Mohr's circles of biaxial and triaxial tests on Berea sandstone.

where a = half-crack length, w = width, k = stress ratio σ_1/σ_3, and Y is a geometric factor also dependent on friction for this problem. It was assumed that $K_I = 0$ and crack propagation was governed by a mode II fracture toughness K_{IIc}. A special interface element was used for the crack surface so that penetration was prohibited. Frictional sliding between the crack faces was allowed with a friction coefficient of 0.67, as determined by a direct shear tests on two saw-cut surfaces. The geometric factor $Y=0.208-0.042 \alpha+0.172 \alpha^2-0.117 \alpha^3+0.025\alpha^4$, a/w=$\alpha$, was determined based on this value of friction and a crack angle of 77°. As expected, crack growth within this geometry was associated with unstable growth.

The loading condition k=13.1 corresponded to Figure 5b with an assumed crack length of 27 mm, shown as the solid line in the figure. The stress intensity factor was calculated to be 3 MPa√m. For Figure 5c with an assumed crack length of 48 mm (also shown as the solid line in the figure) and k=9.66, K_{II} was again found to be 3 MPa√m. Assuming crack propagation only, the controlled portion of the post-peak regime would be dictated by an energy balance, such that $K_{II}=K_{IIc}$, so it seems possible that a crack could have propagated in shear.

At some point, tensile rupture occurred and the kinked region was formed. The condition of $K_I=0$ during shear fracture could have been associated with a closing of the crack during mode II propagation, and at some point additional closure was not possible and tensile fracture was the preferential failure mode.

However, the analysis of only two deformation states is not sufficient to draw a definitive conclusion of shear fracture being responsible for the primary feature. The analysis for the propagation of the primary feature was based on the assumption that shear fracture and not local tensile failure was the controlling mechanism. This has not been supported by other research. In addition, a band of dilatant material surrounding the primary feature is consistent with the notion of a shear band. Further work is on going to investigate the development of the primary feature using AE and digital imaging techniques.

5 CONCLUDING REMARKS

From a plane strain compression test on Berea sandstone at low confinement, the failure plane was characterized by two distinct failure modes. The primary feature appeared to be related to shear-type deformation, with crushed grains and a dilatant zone. Further work is needed to identify the failure mechanism. A shear fracture would not be associated with a band of finite thickness, although a process zone may be responsible for this effect. The second region was absent of shear related features, implying a tensile mode of failure. Furthermore, the transition between the two regions was located where the inclination of the failure surface changed. The termination of the main fracture was observed within the transition zone.

Acknowledgements- Partial support was provided from NSF grant number CMS-0070062.

REFERENCES

Arthur, J.R.F., Dunstan T., Al-Ani Q.A. & Assadi, A. 1977. Plastic deformation and failure in granular media. *Geotechnique*. 27:53-74.

Aydin, A. 1978. Small faults formed as deformation bands in sandstone. *J. Geophys. Res.* 78:2403-2417.

Bernabe, Y. & Brace, W.F. 1996. Deformation and fracture of Berea sandstone. *Am. Geophys. Un. Geophys. Monogr.* 56:91-101.

Besuelle, P., Desrues, J. & Raynaud, S. 2000. Experiemental characterization of the localization phenomenon inside a Vosges sandstone in a triaxial cell. *Int. J. Rock Mech. Min. Sci.* 37:1223-1237.

Dunn, D.E., LaFountain, L.J. & Jackson, R.E. 1973. Porosity dependence and mechanism of brittle fracture in sandstone. *J. Geophys. Res.* 78:2403-2417.

FRANC2D. 1995. *A two dimensional crack propagation simulator.* Cornell Fracture Group. http://www.cfg.cornell.edu.

Labuz, J.F. & Bridel, J.M. 1993. Reducing frictional constraint on compression testing through lubrication. *Int. J. Rock Mech. Min. Sci. Geomech. Abstr.* 30:451-455.

Labuz, J.F., Dai S. –T. & Papamichos. 1996. Plane-strain compression of rock-like materials. *Int. J. Rock Mech. Min. Sci. Geomech. Abstr.* 33:573-584.

Menedez, B., Wenlu, Z. & Wong T. –F. 1996. Micromechanics of brittle faulting and cataclastic flow in Berea sandstone. *J. Struct. Geol.* 18:1-16.

Needleman, A. & Ortiz M. 1991. Effects of boundaries and interfaces on shear-band localization. *Int. J. Solids Structures.* 28:859:877.

Ord, A., Vardoulakis, I. & Kajewski, R. 1991. Shear band formation in Gosford sandstone. *Int. J. Rock Mech. Min. Sci. Abstr.* 28:397-409.

Vardoulakis, I. 1980. Shear-band inclination and shear modulus of sand in biaxial tests. *Int. J. Num. Anal. Meth. Geomech.* 4:103-119.

Vardoulakis, I. & Sulem, J. 1995. *Bifurcation Analysis in Geomechanics.* Blackie Academic and Professional, Glasgow.

5. Non-local models and coupled effects

Bifurcations & Instabilities in Geomechanics, – Labuz & Drescher (eds.)
© 2003 Swets & Zeitlinger, Lisse, ISBN 90 5809 563 0

On the continuous and discontinuous approaches for simulating localized damage

Z. Chen & W. Hu
Department of Civil and Environmental Engineering
University of Missouri-Columbia, Columbia, MO 65211-2200, USA

ABSTRACT: Two different approaches exist for simulating the evolution of localized damage, i.e. continuous and discontinuous ones. Representing a hyperbolic-to-elliptic transition with a parabolic equation, damage diffusion laws have recently been proposed to develop a robust numerical procedure without invoking higher order terms in the single continuous governing differential equation. In this paper, the relationship among rate-dependency, strain gradient and damage diffusion is explored to demonstrate that all these continuous approaches are of higher orders in space and/or time from a mathematical viewpoint although a single higher order equation could be decomposed into a set of 2nd order equations governing different problem domains. It appears that a combined rate-dependent damage and decohesion approach could be sound in physics and efficient in computation if a discontinuous bifurcation analysis is performed to bridge the gap between the continuous and discontinuous approaches. Sample problems are considered to illustrate the potential of the proposed approach in simulating the evolution of impact failure.

1 INTRODUCTION

As can be found from the open literature, two different kinds of approaches have been proposed over the last twenty years to model and simulate the evolution of localized material failure, namely, continuous and discontinuous ones. Decohesion and fracture mechanics models are representative of discontinuous approaches, in which strong discontinuities are introduced into a continuum body such that the governing differential equation is well-posed for given boundary and/or initial data. On the other hand, nonlocal (integral or strain gradient) models, Cosserat continuum models and rate-dependent models are among the continuous approaches proposed to regularize the localization problems, in which the higher order terms in space and/or time are introduced into the strain-stress relations so that the mathematical model is well-posed in a higher order sense for given boundary and/or initial data. Usually, only weak discontinuities in the kinematical field variables are allowed in the continuous approaches; i.e. the continuity of displacement field must hold in the continuum during the failure evolution.

If a continuous approach is of interest, the use of higher order terms in space makes it difficult to perform large-scale computer simulation, due to the limitation of current computational capabilities. As can be found by reviewing the existing nonlocal models, the nonlocal terms are usually included in the limit surface so that a single higher order governing equation would appear in the problem domain. As shown in the previous research (Chen & Sulsky 1995), the evolution of localization might be equally well characterized by the formation and propagation of a moving material surface of discontinuity. An attempt has been made to investigate the use of the jump forms of conservation laws in defining the moving material surface (Chen 1996). By taking the initial point of localization as that point where the type of the governing differential equations changes, a moving material surface of discontinuity can be defined through the jump forms of conservation laws across the surface. Because the transition from a hyperbolic equation to an elliptic one could be represented by a parabolic (diffusion) one, an analytical solution has been obtained for a dynamic softening bar with the use of a similarity method for the transition involving a weak discontinuity (Xin & Chen 2000). Recently, several damage diffusion laws have been proposed to simulate the evolution of localized brittle failure (Chen 2000, Chen & Xin 1999, Chen et al. 2000, Vardoulakis & Papamichos 2000). The use of

& Xin 1999, Chen et al. 2000, Vardoulakis & Papamichos 2000). The use of different governing differential equations in different problem domains makes it possible to replace the single higher order equation with lower order equations so that parallel computing might be used for the large-scale simulation of localization problems.

In this paper, the relationship among rate-dependency, strain gradient and damage diffusion is explored to demonstrate that all these continuous approaches are of higher orders in space and/or time from a mathematical viewpoint although a single higher order equation could be decomposed into a set of 2nd order equations governing different problem domains. It appears that a combined rate-dependent local damage and decohesion approach could be sound in physics and efficient in computation if a discontinuous bifurcation analysis is performed to bridge the gap between the continuous and discontinuous approaches. Sample problems are considered to illustrate the potential of the proposed approach in simulating the evolution of brittle failure under impact loading conditions.

2 THE RELATIONSHIPS AMONG THE REGULARIZING APPROACHES OF LOCALIZATION

For the purpose of simplicity, consider a bar with one end being fixed. A tensile step force is applied at the other end so that localization would occur at the fixed end. The equations of motion and continuity can be written as

$$\frac{\partial \dot{s}}{\partial x} = \rho \frac{\partial^2 v}{\partial t^2} \tag{1}$$

and

$$\dot{e} = \frac{\partial v}{\partial x} \tag{2}$$

where s denotes stress, e strain, v velocity, t time and ρ mass density. The conventional higher-order constitutive model consists of

$$\dot{e} = \dot{e}_e + \dot{e}_i = \frac{\dot{s}}{E} + \dot{e}_i \tag{3.1}$$

$$f\left(s, \dot{s}, I, \nabla I, \nabla^2 I\right) = 0 \tag{3.2}$$

$$\dot{I} = m_0 \dot{e}_i \tag{3.3}$$

in which the total strain rate is decomposed into the rates of elastic and inelastic components, E is the Young's modulus, and the internal state variable I is related to the inelastic strain rate through Eq. (3.3) with m_0 being a model parameter. As can be seen from Eq. (3.2), the higher order terms in space and/or time are included in the limit surface so that a single higher-order governing equation would appear in the problem domain, with the use of the inelastic consistency condition, namely

$$\dot{s} + m_1 \ddot{s} + m_2 \dot{I} + m_3 \nabla \dot{I} + m_4 \nabla^2 \dot{I} = 0 \tag{4}$$

In Eq. (4), $\dfrac{\partial f}{\partial s} = 1$, $\dfrac{\partial f}{\partial \dot{s}} = m_1$, $\dfrac{\partial f}{\partial I} = m_2$, $\dfrac{\partial f}{\partial(\nabla I)} = m_3$ and $\dfrac{\partial f}{\partial(\nabla^2 I)} = m_4$ have been used to simplify the representation.

If a simple diffusion model is coupled with a local limit surface, the constitutive equations consist of

$$\dot{e} = \dot{e}_e + \dot{e}_i = \frac{\dot{s}}{E} + \dot{e}_i \qquad (5.1)$$

$$f(s, \dot{s}, I) = 0 \qquad (5.2)$$

$$\dot{I} = d\nabla^2 e_i \qquad (5.3)$$

where d represents the diffusion coefficient. The corresponding inelastic consistency condition yields

$$\dot{s} + m_1\ddot{s} + m_2\dot{I} = \dot{s} + m_1\ddot{s} + m_2(d\nabla^2 e_i) = 0 \qquad (6)$$

As can be seen from Eqs. (4) and (6), the use of rate-dependency, strain gradient and damage diffusion would all result in a higher-order governing differential equation from a mathematical viewpoint although a single higher order equation could be decomposed into a set of 2nd order equations governing different problem domains if the limit surface is local. In other words, the higher order terms in space and/or time must be employed to simulate the evolution of localization if the continuous approaches are used. Mathematically speaking, the addition of higher order terms into the governing equations makes the localization problem well-posed for given boundary and initial data. However, the physics behind the additional boundary data is still not clear.

As can be found from the open literature, many approaches have been proposed to model and simulate the evolution of localization and the subsequent transition from continuous to discontinuous failure modes. However, there are certain kinds of applicability and limitation for different approaches, depending on the scale of the problem and the degrees of discontinuity considered (Bazant & Chen 1997). Since the discontinuous bifurcation identifies the transition from continuous to discontinuous failure modes, it appears that a combined rate-dependent local damage and decohesion approach could be sound in physics and efficient in computation. As a result, the gap between the continuous and discontinuous approaches could be bridged to simulate a complete failure evolution process without invoking higher order terms in space, as illustrated below.

3 RATE-DEPENDENT DAMAGE AND DECOHESION

To estimate stress-wave-induced fracturing, a combined damage/plasticity model has evolved over a number of years, which was primarily applied to the case of rock fragmentation (Chen et al. 2000). Within the loading regime of the model, an isotropic elasticity tensor governs the elastic material behavior; a scalar measure of continuum damage is active through the rate-dependent degradation of the elasticity tensor if the confining pressure $P \geq 0$ (tensile regime); and a pressure-dependent perfectly plastic model is used if $P < 0$ (compressive regime). The evolution equation for rate-dependent tensile damage can be described with the use of the following equations:

$$C_d = \frac{5k}{2}\left(\frac{K_{IC}}{\rho C \dot{\varepsilon}_{max}}\right)^2 \varepsilon_v^m \qquad (7.1)$$

$$D = \frac{16(1 - \bar{v}^2)}{9(1 - 2\bar{v})}C_d \qquad (7.2)$$

$$\bar{K} = (1 - D)K \qquad (7.3)$$

$$\overline{v} = \left(1 - \frac{16}{9}C_d\right)$$

(7.4)

in which C_d is a crack-density parameter, K_{IC} the fracture toughness, ε_v the mean volumetric strain, $\dot{\varepsilon}_{max}$ the maximum volumetric strain rate experienced by the material at fracture, C the uniaxial wave speed $\sqrt{E/\rho}$ with E being Young's modulus, and D a single damage parameter. Also, K and v are the original bulk modulus and Poisson's ratio, respectively, for the undamaged material, and the barred quantities represent the corresponding parameters of the damaged material. The model parameters k and m can be determined by using the fracture stress versus strain rate curve. With the use of

$$f_1(\overline{v}) = \frac{1 - \overline{v}^2}{1 - 2\overline{v}}$$

(8)

and

$$f_2(\overline{v}) = \frac{2(1 - \overline{v} + \overline{v}^2)}{(1 - 2\overline{v})^2}$$

(9)

the rate forms of Eqs. (7) can be found to be

$$\dot{C}_d = \frac{5km}{2}\left(\frac{K_{IC}}{\rho C\dot{\varepsilon}_{max}}\right)^2 \varepsilon_v^{m-1}\dot{\varepsilon}_v = F_1(\varepsilon_v)\dot{\varepsilon}_v$$

(10.1)

$$\dot{D} = \frac{16}{9}\left[f_1(\overline{v}) - \frac{16}{9}vC_d f_2(\overline{v})\right]\dot{C}_d = F_2(\varepsilon_v)\dot{\varepsilon}_v$$

(10.2)

$$\overline{\dot{K}} = -K\dot{D} = -KF_2(\varepsilon_v)\dot{\varepsilon}_v$$

(10.3)

$$\overline{\dot{v}} = -\frac{16}{9}v\dot{C}_d = -\frac{16}{9}vF_1(\varepsilon_v)\dot{\varepsilon}_v$$

(10.4)

with F_i denoting functionals. As can be seen, the damage evolution can be determined for given $\dot{\varepsilon}_v$ based on the loading/unloading criterion, and a one-step model solver can be designed for large-scale computer simulation.

To obtain a continuum tangent stiffness tensor for the rate-dependent tensile damage, introduce the spherical, $\mathbf{P}^s$, and deviatoric, $\mathbf{P}^d$, orthogonal projections so that the stress-strain relation can be written as

$$\mathbf{s} = \mathbf{E}^{ed} : \mathbf{e} = (3\overline{K}\mathbf{P}^s + 2\overline{G}\mathbf{P}^d) : \mathbf{e}$$

(11)

with $\mathbf{E}^{ed}$ being the elastodamage secant stiffness tensor, and

$$\overline{G} = \frac{3\overline{K}(1 - 2\overline{v})}{2(1 + \overline{v})}$$

(12)

It follows from Eq. (12) that

$$\overline{\dot{G}} = \frac{3\left[(1 - \overline{v} - 2\overline{v}^2)\overline{\dot{K}} - 3\overline{K}\dot{\overline{v}}\right]}{2(1 + \overline{v})^2} = F_3(\varepsilon_v)\dot{\varepsilon}_v$$

(13)

The rate form of Eq. (11) then takes the form of

$$\dot{\mathbf{s}} = \left(3\dot{\overline{K}}\mathbf{P^s} + 2\dot{\overline{G}}\mathbf{P^d}\right) : \mathbf{e} + \left(3\overline{K}\mathbf{P^s} + 2\overline{G}\mathbf{P^d}\right) : \dot{\mathbf{e}} \tag{14}$$

with the tangent stiffness tensor for the rate-dependent tensile damage being given by

$$\mathbf{T}_t^{ed} = \left[-3KF_2(\varepsilon_v)\mathbf{P^s} + 2F_3(\varepsilon_v)\mathbf{P^d}\right] : \mathbf{e} \otimes \frac{\mathbf{i}}{3} + \left(3\overline{K}\mathbf{P^s} + 2\overline{G}\mathbf{P^d}\right) \tag{15}$$

in which $\mathbf{i}$ is the second order identity tensor. As can be seen, the tangent stiffness tensor is not symmetric although the secant stiffness tensor is symmetric. The limit state can be identified by performing the eigenanalysis of the tangent stiffness tensor. However, the discontinuous bifurcation, which results in the transition from continuous to discontinuous failure modes, might occur before the limit state is reached because the tangent stiffness tensor in the tensile regime is not symmetric as illustrated in Figure 1. To identify the transition from continuous to discontinuous failure modes, and the corresponding normal to the surface of discontinuity, therefore, the bifurcation analysis of the acoustic tensor obtained from Eq. (15) must be performed. As shown in the previous work (Chen et al. 2001), the failure angle is rate-independent although the transition level is rate-dependent for the tensile damage model.

After the discontinuous bifurcation occurs, a discrete rate-dependent constitutive model is developed here to predict decohesion or separation of continuum, based on the previous work (Schreyer et al. 1999, Xu & Needleman 1994) in which a discontinuous bifurcation analysis was not performed. With the proposed computational approach, a cohesive surface constitutive relation between the traction and displacement jump is used to predict the evolution of micro-cracking, and a fracture-mechanics-based model can then be employed to trace the crack propagation if a macrocrack occurs. Since the location and the orientation of the cohesive surface is determined via the bifurcation analysis, the mesh-independent (objective) results could be obtained.

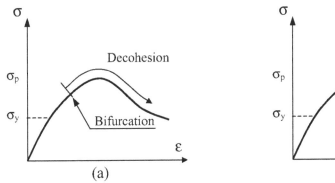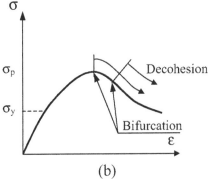

(a)   (b)

Figure 1: Discontinuous bifurcation might occur (a) before the limit state, or (b) at the limit state or after the limit state. After the bifurcation is identified, decohesion will be active.

As shown in Figure 2 for two-dimensional cases, $\mathbf{n}$ and $\mathbf{t}$ denote the unit normal and tangent vectors to the cohesive surface, respectively. To determine the constitutive relation between the traction $\boldsymbol{\tau}$ and decohesion (displacement jump) $\mathbf{u}^d$, the following equations, which satisfy the thermodynamic restrictions, must be solved simultaneously for a given strain rate:

$$\dot{\mathbf{s}} = \mathbf{E}^{ed} : \left(\dot{\mathbf{e}} - \dot{\mathbf{e}}^d\right) \qquad \text{Continuum Elastodamage} \tag{16.1}$$

$$\dot{\boldsymbol{\tau}} = \dot{\mathbf{s}} \cdot \mathbf{n} \qquad \text{Traction Equilibrium} \tag{16.2}$$

$$\dot{\mathbf{u}}^d = \dot{\lambda}\mathbf{m} \qquad \text{Evolution of Decohesion} \tag{16.3}$$

$$\dot{\mathbf{e}}^d = \frac{\dot{\lambda}}{2L_e}(\mathbf{n} \otimes \mathbf{m} + \mathbf{m} \otimes \mathbf{n}) \qquad \text{Decohesion Strain} \qquad (16.4)$$

$$F^d = \tau^e - U_0(1 - \lambda^q) = 0 \qquad \text{Consistency Condition} \qquad (16.5)$$

in which λ is a dimensionless monotonically increasing variable parametrizing the evolution of decohesion, and L_e is the effective length representing the ratio of the volume to the area of the decohesion within a material element. For the purpose of simplicity, an associated evolution equation is employed, namely

$$\mathbf{m} = \bar{u}_0 \frac{\mathbf{A_d} \cdot \tau}{(\tau \cdot \mathbf{A}_d \cdot \tau)^{1/2}} \qquad (17)$$

so that the effective traction takes the form of

$$\tau^e = \tau \cdot \mathbf{m} = \bar{u}_0 (\tau \cdot \mathbf{A_d} \cdot \tau)^{1/2} \qquad (18)$$

with the reference surface energy U_0 being the product of the reference decohesion scalar $\bar{u}_0$ and corresponding scalar traction $\bar{\tau}_0$. The components of the positive definite tensor of material parameters, $\mathbf{A_d}$, with respect to the $\mathbf{n} - \mathbf{t}$ basis are given by

$$[\mathbf{A_d}] = \bar{\tau}_0^2 \begin{bmatrix} \dfrac{1}{\tau_{np}^2} & 0 \\ 0 & \dfrac{1}{\tau_{tp}^2} \end{bmatrix} \qquad (19)$$

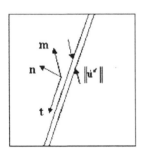

Figure 2: A two-dimensional material element with decohesion.

At the initiation of decohesion, it follows from Eqs. (16.5), (18) and (19) with $\lambda = 0$ that

$$\frac{\tau_{nb}^2}{\tau_{np}^2} + \frac{\tau_{tb}^2}{\tau_{tp}^2} = 1 \qquad (20)$$

where the normal and tangential tractions, τ_{nb} and τ_{tb}, are determined from the discontinuous bifurcation analysis, and depend on both the strain state and strain rate, i.e.

$$\tau_{nb} = \tau_{nb}(\mathbf{e}, \dot{\mathbf{e}}) \tag{21.1}$$

$$\tau_{tb} = \tau_{tb}(\mathbf{e}, \dot{\mathbf{e}}) \tag{21.2}$$

By letting $C_m = \dfrac{\tau_{tp}}{\tau_{np}}$, different failure modes can be simulated by using different values of C_m and Eq. (20). For example, mode I failure dominates if $C_m = 10$, while mode II failure dominates if $C_m = 0.1$. Mixed failure modes could be simulated by using $C_m = 1$, depending on the critical state. As can be seen from the above formulations, the discrete model parameters to be determined from the experiments are U_0, q and C_m if the choice of $\tau_{np} = \overline{\tau}_0$ is made. The relation between the traction and decohesion can be adjusted via changing the value of q, as shown in Figure 3.

To design an incremental forward-integration scheme without iteration for the discrete constitutive model, a Taylor expansion of F^d about the trial state can be obtained by enforcing $F^d = 0$ to the order of $(\Delta\lambda)^2$, namely

$$F^d = \left.\frac{\partial F^d}{\partial\lambda}\right|_{tr}\Delta\lambda + \left[\tau^e - U_0(1 - \lambda^q)\right]_{tr} + O(\Delta\lambda)^2 = 0 \tag{22}$$

For a given total strain increment, assume first that no decohesion occurs to obtain a trial stress and traction, and then determine the value of the decohesion function, $\left.F^d\right|_{tr}$, for this trial traction and the existing value of λ. If $\left.F^d\right|_{tr} \leq 0$, the step is elastodamaging with no additional decohesion. If not, $\Delta\lambda$ is determined from Eq. (22), and the variables related to decohesion must be incremented.

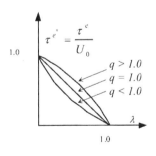

Figure 3: Decohesion relation in terms of λ.

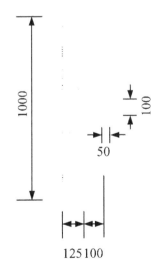

Figure 4. The initial configuration of a plane-strain impact problem. Unit (mm)

4 IMPLEMENTATION OF CONSTITUTIVE MODELS AND DEMONSTRATION

4.1 The material point method

To accommodate the multi-scale discontinuities involved in structural failure, a robust simulation tool is a necessity without invoking a fixed mesh connectivity. As one of innovative spatial discretization methods, the Material Point Method (MPM) is an extension to solid mechanics problems of a hydrodynamics code which, in turn, evolved from the Particle-in-Cell Method. The motivation of the development was to simulate those problems with history-dependent state variables, such as contact/impact, penetration/perforation and solid-fluid interaction, without invoking master/slave nodes and remeshing. The essential idea is to take advantages of both Eulerian and Lagrangian methods. Although the MPM is still under development, sample calculations have demonstrated the robustness and potential of this method for the challenging problems of current interests, as shown in the representative references (Chen et al. 2002, Sulsky et al. 1994, Zhou et al. 1999). The essential idea of the MPM is summarized as follows.

The MPM discretizes a continuum body with the use of a finite set of N_p material points in the original configuration that are tracked throughout the deformation process. Let $\mathbf{x}_p^t$ ($p = 1, 2, ..., N_p$) denote the current position of material point p at time t. Each material point at time t has an associated mass M_p, density ρ_p^t, velocity $\mathbf{v}_p^t$, Cauchy stress tensor $\mathbf{s}_p^t$, strain $\mathbf{e}_p^t$, and any other internal state variables necessary for constitutive modeling. Thus, these material points provide a Lagrangian description of the continuum body. Since each material point contains a fixed amount of mass for all time, the conservation of mass is automatically satisfied. At each time step, the information from the material points is mapped to a background computational mesh (grid). This mesh covers the computational domain of interest, and is chosen for computational convenience. After the information is mapped from the material points to the mesh nodes, the discrete equations of the conservation of momentum can be solved on the mesh nodes. The weak form of the conservation of momentum can be found, based on the standard procedure used in the Finite Element Method, to be

$$\int_{\Omega} \rho \mathbf{w} \cdot \mathbf{a} d\Omega = -\int_{\Omega} \rho \mathbf{s}^s : \nabla \mathbf{w} d\Omega + \int_{S^c} \rho \mathbf{c}^s \cdot \mathbf{w} dS + \int_{\Omega} \rho \mathbf{w} \cdot \mathbf{b} d\Omega \tag{23}$$

in which $\mathbf{w}$ denotes the test function, $\mathbf{a}$ is the acceleration, $\mathbf{s}^s$ is the specific stress (i.e. stress divided by mass density), $\mathbf{c}^s$ is the specific traction vector (i.e. traction divided by mass density), $\mathbf{b}$ is the specific body force, Ω is the current configuration of the continuum, S^c is that part of the boundary with a prescribed traction. The test function $\mathbf{w}$ is assumed to be zero on the boundary with a prescribed displacement. Since the whole continuum body is described with the use of a finite set of material points (mass elements), the mass density term can be written as

$$\rho(\mathbf{x}, t) = \sum_{p=1}^{N_p} M_p \delta(\mathbf{x} - \mathbf{x}_p^t) \tag{24}$$

where δ is the Dirac delta function with dimension of the inverse of volume. The substitution of Eq. (24) into Eq. (23) converts the integrals to the sums of quantities evaluated at the material points, namely

$$\sum_{p=1}^{N_p} M_p \left[\mathbf{w}(\mathbf{x}_\mathbf{p}^t, t) \cdot \mathbf{a}(\mathbf{x}_\mathbf{p}^t, t) \right]$$

$$= \sum_{p=1}^{N_p} M_p \left[-\mathbf{s}^\mathbf{s}(\mathbf{x}_\mathbf{p}^t, t) : \nabla \mathbf{w} \Big|_{\mathbf{x}_\mathbf{p}^t} + \mathbf{w}(\mathbf{x}_\mathbf{p}^t, t) \cdot \mathbf{c}^\mathbf{s}(\mathbf{x}_\mathbf{p}^t, t) h^{-1} + \mathbf{w}(\mathbf{x}_\mathbf{p}^t, t) \cdot \mathbf{b}(\mathbf{x}_\mathbf{p}^t, t) \right] \qquad (25)$$

with h being the thickness of the boundary layer. As can be seen from Eq. (25), the interactions among different material points are reflected only through the gradient terms. Because there is no fixed mesh connectivity in the MPM, impact, localization and the transition from continuous to discontinuous failure modes can be simulated without the difficulties associated with remeshing and master/slave nodes, as demonstrated by the following sample problems.

4.2 Demonstration

The computational domain for the model problem is illustrated in Figure 4 in which a concrete wall is subjected to a high-velocity steel flyer. As described by Zukas (Zukas 1994), high velocity impact at impact velocities of 0.5~2.0 km/s can be complicated by the presence of new material interface. Here, we focus on demonstrating the evolution of localized damage in the tensile regime of concrete wall, in which the continuous description of damage is switched to the discontinuous one through the bifurcation analysis.

The following assumptions are made for the purpose of simplicity:
- No crack closure effect is taken into account.
- No strain rate effect is considered in the compressive regime.
- No bifurcation analysis is performed in the compressive regime.

The size of the entire computational domain is 1.5m x 1.5m, which is discretized with a 60 x 60 uniform grid. Initially, there are four material points in each grid cell. This is a plane-strain problem with free boundary. An explicit time integration scheme is used with the time step satisfying the stability criteria.

An elastic-perfectly plastic von Mises model with an associated flow rule is used to describe the steel flyer with Young's modulus $E = 200.0 GPa$, density $\rho = 7850 kg/m^3$ and Possion's ratio $\upsilon = 0.29$. The yield strength of steel is $500 MPa$. For the concrete wall, the model parameters are assigned the following values: Young's modulus $E = 25.0 GPa$, density $\rho = 2320 kg/m^3$, and Possion's ratio $\upsilon = 0.15$, with the fracture toughness $K_{IC} = 1.0 MPa\sqrt{m}$. Based on the work (Taylor et al. 1986), the model parameters m and k for concrete are chosen to be 6 and $5 \times 10^{26} 1/m^3$ in the tensile regime, respectively.

An elastic-perfectly plastic Drucker-Prager model with a non-associated flow rule is used in the compressive regime of the concrete wall. Note that the rate-dependent damage as described in the previous section is active in the tensile regime of the concrete wall before the discontinuous bifurcation occurs. At each time step, the bifurcation analysis of acoustic tensor must be performed until the localization line in the strain space of Mohr osculates the major principal circle of strain (Chen et al. 2001). As soon as the transition from continuous to discontinuous failure modes is identified, the decohesion model is active with mode I failure being dominant. The failure initiation value, τ_{np}, is found to be the largest principal stress based on the bifurcation analysis. C_m is chosen to be 10, and $\overline{\tau}_0 = \tau_{np}$ is used for the mode I failure. The reference decohesion scalar, $\overline{u}_0$, is assigned to be 6×10^{-4} m. The parameter q is chosen to be 1.0, which provides a linear relation between the traction and decohesion. The main flowchart of implementation of constitutive models for the impact problem is illustrated in Figure 5.

227

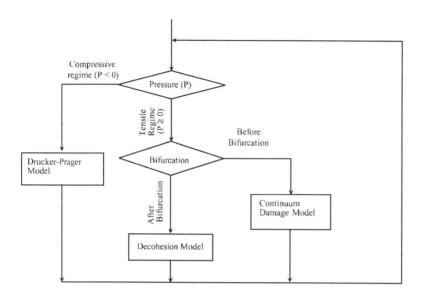

Figure 5: The main flowchart of implementation of constitutive models for the impact problem.

The failure patterns of the concrete wall due to the impact of steel flyer at time 0.442 *ms* are shown in Figures 6a, 6b and 6c for different impact velocities. Initially the concrete wall is at rest. As the flyer approaches the concrete target, contact/impact between the flyer and wall produces acceleration of the concrete wall and deceleration of the steel flyer. Through the interaction among individual material particles, the failure evolution is spread toward the edges of the concrete wall. The steel flyer therefore becomes 'submerged' among the fragments of concrete wall. It can be found by comparing Figures 6a, 6b and 6c that the failure pattern of the concrete wall depends on the impact velocity. The higher the impact velocity, the more significant velocity the fragments acquire. The shape of the concrete wall has therefore changed more dramatically. With the increase of impact velocity, the breadth of the impact hole increases. Note that the normal to the decohesion surface at the center point of the concrete wall is parallel to the x coordinate based on the bifurcation analysis. Also, as can be observed from Figures 7a, 7b and 7c, the impact induces significant material failure close to the middle section of the concrete wall. The higher the impact velocity, the more localized damage gradient happens. Coupled with the presence of multiple wave reverberations due to geometric effects, the most intense deformations occur within 2-3 times of the characteristic dimensions of the flyer. Within this region, strain rates in excess of 10,000/s are not uncommon, as are large strain, and pressures well in excess of the material strength. In the simulation, we did not make use of the initial symmetry of the problem. The problem as a whole was analyzed and motion of all material points was traced. It is worth noting that the initial symmetry of the system is in general preserved although localized loss of symmetry is observed. As can be seen, the MPM does not exhibit the finite element pathologies associated with distorted meshes and instabilities, and the MPM does not exhibit the orientation effect often seen with finite elements when discontinuities are allowed to propagate at various angles to the mesh sizes. Hence, characteristics of impact, localization and the rate-dependent transition from continuous to discontinuous failure modes have been qualitatively predicted by the proposed procedure.

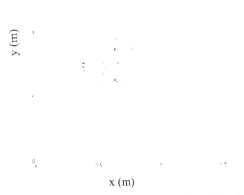

Figure 6a: The failure pattern with impact velocity being 1000 m/s.

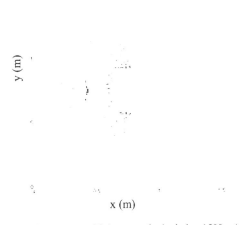

Figure 6b: The failure pattern with impact velocity being 1500 m/s.

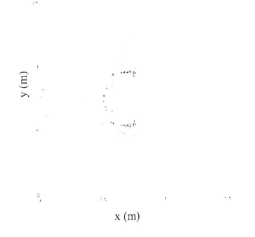

Figure 6c: The failure pattern with impact velocity being 2000 m/s.

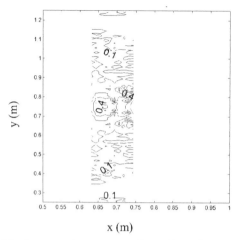

Figure 7a: The damage contour corresponding to Figure 6a.

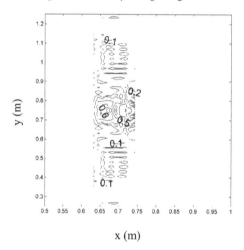

Figure 7b: The damage contours corresponding to Figure 6b.

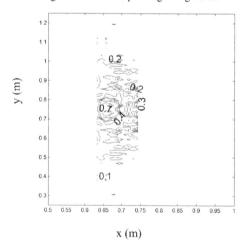

Figure 7c: The damage contours corresponding to Figure 6c.

5 CONCLUSIONS

In this paper, the relationship among rate-dependency, strain gradient and damage diffusion has been explored to demonstrate that all these continuous approaches are of higher orders in space and/or time from a mathematical viewpoint although a single higher order equation could be decomposed into a set of 2nd order equations governing different problem domains. It appears that a combined rate-dependent local damage and decohesion approach could be sound in physics and efficient in computation if a discontinuous bifurcation analysis is performed to bridge the gap between the continuous and discontinuous approaches. To accommodate the multi-scale discontinuities involved in dynamic structural failure, the MPM appears to be a robust spatial discretization tool. Future work will focus on verification and validation of the proposed model-based simulation procedure.

ACKNOWLEDGEMENT

This research was sponsored in part by the National Science Foundation with Dr. K. Chong being program director.

REFERENCES

Bazant, Z.P. & Chen, E.P. 1997. Scaling of Structural Failure. *Applied Mechanics Reviews* 50: 593-627.

Chen, Z. 1996. Continuous and Discontinuous Failure Modes. *Journal of Engineering Mechanics* 122(1): 80-82.

Chen, Z. 2000. Simulating the Evolution of Localization Based on the Diffusion of Damage. *International Journal of Solids and Structures* 37: 7465-7479.

Chen, Z., Deng, M. & Chen, E.P. 2001. On the Rate-Dependent Transition from Tensile Damage to Discrete Fracture in Dynamic Brittle Failure. *Theoretical and Applied Fracture Mechanics* 35(3): 229-235.

Chen, Z., Hu, W. & Chen, E.P. 2000. Simulation of Dynamic Failure Evolution in Brittle Solids without Using Nonlocal Terms in the Strain-Stress Space. *Computer Modeling in Engineering & Sciences* 1(4): 101-106.

Chen, Z., Hu, W., Shen, L., Xin, X. & Brannon, R. 2002. An Evaluation of the MPM for Simulating Dynamic Failure with Damage Diffusion. *Engineering Fracture Mechanics* 69: 1873-1890.

Chen, Z. & Sulsky, D. 1995. A Partitioned-Modeling Approach with Moving Jump Conditions for Localization. *International Journal of Solids and Structures* 32: 1893-1905.

Chen, Z. & Xin, X. 1999. An Analytical and Numerical Study of Failure Waves. *International Journal of Solids and Structures* 36: 3977-3991.

Schreyer, H.L., Sulsky, D.L. & Zhou S.-J. 1999. Modeling Material Failure as a Strong Discontinuity with the Material Point Method. In G. Pijaudier-Cabot et al. (eds), *Mechanics of Quasi-Brittle Materials and Structures: A Volume in Honor of Professor Zdenek P. Bazant's 60th Birthday*: 307-329. Paris: Hermes Science Publications.

Sulsky, D., Chen, Z. & Schreyer, H.L. 1994. A Particle Method for History-Dependent Materials. *Computer Methods in Applied Mechanics and Engineering* 118: 179-196.

Taylor, L.M., Chen, E.P. & Kuszmaul, J.S. 1986. Microcrack-Induced Damage Accumulation in Brittle Rock under Dynamic Loading. *Computer Methods in Applied Mechanics and Engineering* 55: 301-320.

Vardoulakis, I. & Papamichos, E. 2000. Anisotropic Damage Diffusion. In J. Katsikadelis et al. (eds), *Research Advances in Applied Mechanics* :328-340. Athanasopoulos-Papdamis Publishers.

Xin, X. & Chen, Z. 2000. An Analytical Solution with Local Elastoplastic Models for the Evolution of Dynamic Softening. *International Journal of Solids and Structures* 37: 5855-5872.

Xu, X.P. & Needleman, A. 1994. Numerical Simulations of Fast Crack Growth in Brittle Solids. *Journal of the Mechanics and Physics of Solids* 42: 1397-1434.

Zhou, S.J., Stormont, J. & Chen, Z. 1999. Simulation of Geomembrane Response to Subsidence by Using the Material Point Method. *International Journal for Numerical and Analytical Methods in Geomechanics* 23: 1977-1994.

Zukas, J.A. 1994. Numerical Simulation of High Rate Behavior. In C.A. Brebbia & V. Sanchez-Galvez (eds), *Shock and Impact on Structures* :1-26. Southampton: Computational Mechanics Publication.

Bifurcations & Instabilities in Geomechanics, – Labuz & Drescher (eds.)
© 2003 Swets & Zeitlinger, Lisse, ISBN 90 5809 563 0

Instability and strain localization of water-saturated clay based on an elasto-viscoplastic model

F. Oka, Y. Higo, & S. Kimoto
Department of Civil Engineering, Kyoto University, Kyoto 606-8501, Japan

ABSTRACT: The aim of the present paper is to study the instability and strain localization of water saturated clay. In particular, effects of dilatancy and strain rate are discussed using an elasto-viscoplastic constitutive model. The model is based on the non-linear kinematic hardening theory and a Chaboche type of viscoplasticity model. The elasto-viscoplastic model for both normally consolidated and overconsolidated clays can address both negative and positive dilatancies. Firstly, the instability of the model under undrained creep conditions is presented in terms of the accelerating creep failure. The analysis shows that clay with positive dilatancy is more unstable than clay with negative dilatancy. Secondly, a finite element analysis of the deformation of water-saturated clay is presented with focus on the numerical results under plane strain conditions. From the present numerical analysis, it is found that both dilatancy and strain rate prominently affect shear strain localization behavior.

1 INTRODUCTION

The problem of strain localization in such geomaterials as soils and rocks has been studied in the context of bifurcation and instability(e.g. Rice, 1976). It has been observed that strain localization appeared as a shear banding in the experiment before failure and post failure behaviors of geomaterials.

The present study deals with the behavior of clay, in which the aspect of rate dependency comes naturally into the modeling. In addition, since the transport of water must be considered in the behavior of water-saturated clay, the problem is formulated within the solid-fluid two-phase theory. Oka et al. (1994, 1995 2000 & 2002) studied strain localization problems pertinent to water-saturated clays using a viscoplastic model. In particular, it was found that strain localization in the shear band of water-saturated clays could be simulated via a finite element analysis using an elasto-viscoplastic model with viscoplastic softening (Oka et al. 1995). However, the model used in the analysis was limited to normally consolidated clay with negative dilatancy.

In the present study, an elasto-viscoplastic model which can describe both negative and positive dilatancy has been developed based on a Chaboche type of viscoplastic theory (Chaboche 1980) and the kinematic hardening rule with viscoplastic softening. At first, the instability of the proposed model is examined under undrained triaxial creep conditions, and then, the effects of dilatancy and strain rate are numerically investigated for both dilatant and contractant clays using the newly developed elasto-viscoplastic constitutive model for both normally and overconsolidated clays.

In the numerical simulation of the shear bands, it has been realized that instability and ill-posedness might be encountered in the problem when using a rate-independent elasto-plastic model in the numerical analysis(Vardoulakis and Aifantis, 1991). In order to retrieve these problems, several methods are taken in the analysis:viscoplastic formulation, non-local formulation of constitutive modeling such as strain gradient theory and discrete model of materials(Mühlhaus and Aifantis, 1991). In addition pore water migration reduces the ill-posedness of the problem(Loret and Prevost, 1991, Schrefler et al., 1996,Oka et al., 2000). In the present analysis, since we adopted an elasto-viscoplastic constitutive model with transport of pore water in the material, numerical instability can be efficiently weakened in the simulation.

2 ELASTO-VISCOPLASTIC MODEL FOR CLAY

Adachi and Oka(1982) proposed an elasto-viscoplastic model for water-saturated normally consolidated clay based on a Perzyna type of viscoplasticity theory and Cam-clay model, has been applied to many practical problems and strain localization problems(Oka et al., 1995). Then an extended elasto-viscoplastic model for both normally-consolidated and overconsolidated clays, based on the kinematic hardening viscoplastic theory proposed by Chaboche (1980), has been developed(Oka, Higo and Kimoto, 2002) The model can be seen as an extension of the overstress type of viscoplastic model developed by Adachi and Oka (1982) for the behavior of normally and quasi-overconsolidated clays. In the present analysis the extended model was used.

In the following, Terzaghi's effective stress concept is used for water-saturated soil because the compressibility of the pore water is effectively small, i.e.

$$\sigma_{ij} = \sigma'_{ij} + u_w \delta_{ij} \tag{1}$$

where σ_{ij} is the total stress tensor, σ'_{ij} is the effective stress tensor, u_w is the pore water pressure, and δ_{ij} is Kronecker's delta. Furthermore, an additive decomposition of the total strain rate into elastic,$\dot{\varepsilon}^e_{ij}$, and viscoplastic,$\dot{\varepsilon}^{vp}_{ij}$, ones is assumed such that

$$\dot{\varepsilon}_{ij} = \dot{\varepsilon}^e_{ij} + \dot{\varepsilon}^{vp}_{ij}. \tag{2}$$

Elastic strain rate $\dot{\varepsilon}^e_{ij}$ is given by a generalized Hooke type of law, i.e.,

$$\dot{\varepsilon}^e_{ij} = \frac{1}{2G}\dot{S}_{ij} + \frac{\kappa}{3(1+e)\sigma'_m}\dot{\sigma}'_m \delta_{ij} \tag{3}$$

where S_{ij} is the deviatoric stress rate tensor, σ'_m is the mean effective stress, G is the elastic shear coefficient, e is the void ratio, κ is the swelling index, and the superimposed dot denotes time differentiation. Swelling index κ is determined by the slope of the volumetric loading-unloading curve of the natural logarithmic scale.

In the model, it is assumed that there is an overconsolidation (OC) boundary surface that delineates the overconsolidation region ($f_b < 0$) from the normal consolidation region ($f_b \geq 0$) (see Figure 1). The overconsolidation boundary surface was introduced to control the shape of the plastic potential function.

In order to describe the volumetric relaxation and/or secondary compression under isotropic stress conditions, it is assumed that the stress state of normally consolidated clay is generally outside of the OC boundary surface and defined as:

$$f_b = \bar{\eta}^* + M_m^* \ln \frac{\sigma'_m}{\sigma'_{mb}} = 0 \tag{4}$$

$$\bar{\eta}^* = \{(\eta^*_{ij} - \eta^*_{ij(0)})(\eta^*_{ij} - \eta^*_{ij(0)})\}^{1/2} , \quad \eta^*_{ij} = \frac{S_{ij}}{\sigma'_m}$$

$$\sigma'_{mb} = \sigma'_{mbi} \exp\left(\frac{1+e}{\lambda - \kappa}\varepsilon^{vp}_{kk}\right) \tag{5}$$

$$\sigma'_{mc} = \sigma'_{mb} \exp\left(\frac{\eta^*_{(0)}}{M_m^*}\right) = \sigma'_{mbi} \exp\left(\frac{1+e}{\lambda - \kappa}\varepsilon^{vp}_{kk}\right) \exp\left(\frac{\eta^*_{(0)}}{M_m^*}\right) \tag{6}$$

$$\eta^*_{(0)} = \sqrt{\eta^*_{ij(0)}\eta^*_{ij(0)}} \quad , \quad \eta^*_{ij(0)} = \frac{S_{ij(0)}}{\sigma'_{m(0)}}$$

in which (0) denotes the state at the end of consolidation, σ'_{mbi} is the initial value of σ'_{mb}, λ is the compression index, and M^*_m is the value of η^* at maximum compression. Compression index λ is determined by the slope of the volumetric loading curve of the natural logarithmic scale. In Figure 1, $\eta_{11(0)}(= \sqrt{3/2}\eta^*_{11(0)})$ stands for the anisotropic consolidation history.

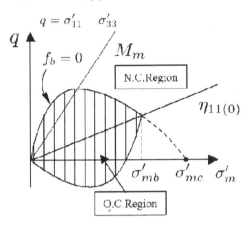

Figure 1 Overconsolidation boundary surface
under the triaxial conditions

A viscoplastic flow rule is given by

$$\dot{\varepsilon}^{vp}_{ij} = C_{ijkl} \langle \Phi_1(f_y) \rangle \Phi_2(\xi) \frac{\partial f_p}{\partial \sigma'_{kl}} \tag{7}$$

$$C_{ijkl} = a\delta_{ij}\delta_{kl} + b(\delta_{ik}\delta_{jl} + \delta_{il}\delta_{jk}) \tag{8}$$

$$C_{o1} = 2b, \ C_{o2} = 3a + 2b$$

in which $<>$ is the MacCauley's bracket; $< x >- x, if \ x > 0,- 0, if \ x \le 0$, C_{01}, C_{02} are viscoplastic parameters, $C_{ijkl} \langle \Phi_1(f_y) \rangle$ denotes a function for strain rate sensitivity, f_y is the yield function, f_p is the plastic potential function, and $\Phi_2(\xi)$ controls the failure state where deviatoric strain becomes infinite.

Based on the experimental results of strain-rate constant triaxial tests(Adachi et al. 1987; Oka et al. 1994), Φ_1 is defined as

$$\Phi_1(f_y) - \exp(m' f_y), \tag{9}$$

where m' is the viscoplastic parameter for a given degree of rate sensitivity.

Yield function f_y with two kinematic hardening parameters, namely, x^*_{ij} and y^*_m, is given by

$$f_y = \bar{\eta}^*_{\chi} + \dot{M}^*|\ln\frac{\sigma'_m}{\sigma'_{ma}} - y^*_m| = 0 \tag{10}$$

$$\bar{\eta}^*_{\chi} = \left\{\left(\eta^*_{ij} - x^*_{ij}\right)\left(\eta^*_{ij} - x^*_{ij}\right)\right\}^{\frac{1}{2}} \quad , \quad \eta^*_{ij} = \frac{S_{ij}}{\sigma'_m}$$

where σ'_{ma} is taken as the initial value of the mean effective stress.

Herein, two strain-hardening parameters are used in the model, namely, x^*_{ij}, which depends on the viscoplastic shear strain rate, and y^*_m, which is related to volumetric viscoplastic strain ε^{vp}_{kk}.

The evolutional equations of x^*_{ij} and y^*_m are given by

$$dx^*_{ij} = B^*_1 \left(A^*_1 de^{vp}_{ij} - x^*_{ij} d\gamma^p \right) \tag{11}$$

$$de^{vp}_{ij} = d\varepsilon^{vp}_{ij} - \frac{1}{3} d\varepsilon^{vp}_{kk} \delta_{ij} \quad , \quad d\gamma^p = \sqrt{de^{vp}_{ij} de^{vp}_{ij}} \tag{12}$$

$$dy^*_m = dy^*_{m1} + dy^*_{m2} \tag{13}$$

$$dy^*_{m1} = B^*_2 \left(A^*_2 d\varepsilon^{vp}_{kk} - y^*_{m1} |d\varepsilon^{vp}_{kk}| \right) \tag{14}$$

$$dy^*_{m2} = \frac{1+e}{\lambda - \kappa} d\varepsilon^{vp}_{kk} \tag{15}$$

where A^*_1, A^*_2, A^*_3, B^*_1, and B^*_2 are material parameters and $A^*_1 (= M^*_f)$ is the value of η^* at the failure state.

A second material function, Φ_2, that is dependent on the internal state variables, is chosen to control the failure state. Basically, it is assumed that Φ_2 becomes infinite at failure (Adachi et al. 1987, 1990), i.e.,

$$\Phi_2(\xi) = 1 + \xi \tag{16}$$

in which ξ is an internal variable.

In general, ξ follows an evolutional equation whose integrated form, satisfying the above-mentioned failure requirement for the internal variable, is given by

$$\xi = \frac{M^*_f \bar{\eta}^{**}_{x(0)}}{G^*_2 \left\{ M^*_f - \frac{\eta^{**}_{mn}(\eta^{**}_{mn} - x^*_{mn})}{\bar{\eta}^{**}_r} \right\}} \tag{17}$$

where G^*_2 is a parameter for the second material function and M^*_f is the value of the stress invariant ratio at failure and

$$\bar{\eta}^{**}_x = \left\{ \left(\eta^{**}_{ij} - x^*_{ij} \right) \left(\eta^{**}_{ij} - x^*_{ij} \right) \right\}^{\frac{1}{2}}, \quad \eta^{**}_{ij} = \frac{S^*_{ij}}{\sigma'^*_m} \tag{18}$$

$$\bar{\eta}^{**}_{x(0)} = \left\{ \left(\eta^{**}_{ij} - x^*_{ij(0)} \right) \left(\eta^{**}_{ij} - x^*_{ij(0)} \right) \right\}^{\frac{1}{2}} \tag{19}$$

where η^{**}_{ij} is the stress history invariant ratio, S^*_{ij} and σ'^*_m are the deviatoric and the mean components of stress history tensor σ^*_{ij}, respectively, and $x^*_{ij(0)}$ denotes the initial value of x^*_{ij}.

Furthermore, stress history tensor σ^*_{ij} is defined as

$$\sigma^*_{ij} = \frac{1}{\tau} \int_0^z \exp\left(-(z - z')/\tau \right) \sigma_{ij}(z') dz' \quad , \quad 0 \le z' < z \tag{20}$$

$$z = \int_0^t dz', \quad dz' = \sqrt{de_{ij} de_{ij}} \tag{21}$$

where t is time, de_{ij} is the increment of deviatoric strain, and τ is a material parameter.

236

Table 1. Material parameters used in the calculations

Parameter		N.C. clay	O.C. clay
Compression index	λ	0.172	0.172
Swelling index	κ	0.054	0.054
Initial void ratio	e_0	0.72	0.72
Initial mean effective stress	σ'_{me}	392 (kPa)	100 (kPa)
Parameter of O.C. boundary surface	σ'_{mbi}	392 (kPa)	392 (kPa)
Coefficient of earth pressure at rest※	K_0	1.0	1.0
Viscoplastic parameter	m	21.5	21.5
Viscoplastic parameter	C_{o1}	4.5×10^{-8}(1/s)	4.5×10^{-8}(1/s)
Viscoplastic parameter	C_{o2}	4.5×10^{-8}(1/s)	4.5×10^{-8}(1/s)
Stress ratio at failure	M_f^*	1.05	1.05
Stress ratio at maximum compression	M_m^*	1.05	1.05
Elastic shear modulus	G	5500 (kPa)	5500 (kPa)
Softening parameter	G_2^*	100	1
Kinematic hardening parameter	B_1^*	0.0	0.5
Kinematic hardening parameter※※	A_2^*	0.0	0.0
Kinematic hardening parameter※※	B_2^*	0.0	0.0
Retardation parameter	τ	1.0×10^{-3}	0.2
Coefficient of permeability※	k	1.54×10^{-6} 1.54×10^{-8} 1.54×10^{-10} (m/s)	1.54×10^{-6} 1.54×10^{-8} 1.54×10^{-10} (m/s)

※ K_0 and coefficient of permeability were used in finite element analysis.
※※ A_2^* and B_2^* were not used in the present analysis.

The stress history tensor was advocated by Oka (1985) and Adachi and Oka (1995). In their theory, both the yield and the hardening functions depend on the stress history rather than on the real stress in order to describe the strain-softening behavior of geomaterials. From the assumption that the second material function, Φ_2, is a function of stress history ratio tensor η_{ij}^{**}, it becomes possible to describe the material behavior in which the stress path can reach a point over the failure line. This type of behavior is dominant for overconsolidated clay as was revealed in the experiments.

The plastic potential is given by Equation (22), which is similar to the yield function, in other words,

$$f_p = \bar{\eta}_x^* + \tilde{M}^* |\ln \frac{\sigma'_m}{\sigma'_{mp}} - y_m^*| = 0 \tag{22}$$

$$\bar{\eta}_x^* = \left\{ \left(\eta_{ij}^* - x_{ij}^*\right) \left(\eta_{ij}^* - x_{ij}^*\right) \right\}^{\frac{1}{2}} \quad , \quad \eta_{ij}^* = \frac{S_{ij}}{\sigma'_m}$$

$$f_b < 0 \quad ; \quad \tilde{M}^* = -\frac{\bar{\eta}^*}{\ln (\sigma'_m / \sigma'_{mc})} \tag{23}$$

$$f_b \geq 0 \quad ; \quad \tilde{M}^* = M_m^* \tag{24}$$

where f_b is the overconsolidation boundary surface given by Equation (4), and σ'_{mp} is a material parameter, herein taken to be equal to the initial value of σ'_m.

Using the flow rule and the plastic potential function, we obtain the deviatori viscoplastic strain rates $\dot{e}_{ij}^{vp}$ and volumetric viscoplastic strain rate $\dot{\varepsilon}_{kk}^{vp}$ as:

$$\dot{e}_{ij}^{vp} = C_{01} \exp \left\{ m' \left(\bar{\eta}_x^* + \tilde{M}^* |\ln \frac{\sigma'_m}{\sigma'_{ma0}} - y_m^*| \right) \right\} \Phi_2 (\xi) \frac{\left(\eta_{ij}^* - x_{ij}^*\right)}{\bar{\eta}_x^*} \tag{25}$$

$$\dot{\varepsilon}_{kk}^{vp} = C_{02} \exp \left\{ m' \left(\bar{\eta}_x^* + \tilde{M}^* |\ln \frac{\sigma'_m}{\sigma'_{ma0}} - y_m^*| \right) \right\} \Phi_2 (\xi) \left\{ \tilde{M}^* - \frac{\eta_{mn}^* (\eta_{mn}^* - x_{mn}^*)}{\bar{\eta}_x^*} \right\} \tag{26}$$

237

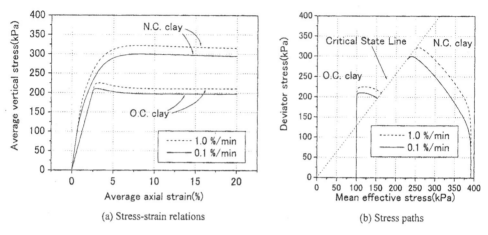

(a) Stress-strain relations   (b) Stress paths

Figure 2 Stress-strain relations and stress paths under undrained triaxial compression conditions

From Eq.(26), it is seen that the sign of the volumetric inelastic strain rate depends on the stress state even inside the overconsolidation boundary surface, because the volumetric strain depends on the value of $\tilde{M}^*$ given by Eqs.(23) and (24), and inside the overconsolidation boundary surface, $\tilde{M}^*$ is a function of σ'_{mc} defined by Eq.(6), which is related to the shape of overconsolidation boundary surface shown in Fig.1.

Figures 2 show the stress-strain relations and the stress paths under undrained triaxial compression conditions. Table 1 lists the sixteen material parameters, including two parameters of the initial conditions, that were used in the analysis in which a linear kinematic hardening equation is assumed for changes in the mean effective stress. Three parameters are different for overconsolidated and normally consolidated clays. This means that the magnitude of strain softening and kinematic hardening depends on the magnitude of overconsolidation. Both the strain rate effect and the strain-softening behavior are observed in Figures 2. From the stress paths, it is seen that the mean effective stress increases due to positive dilatancy for overconsolidated clay, while the mean effective stress decreases due to negative dilatancy for normally consolidated clay.

3 INSTABILITY OF THE MODEL

Oka et al.(1995) studied the instability of the viscoplastic model in terms of undrained creep failure for normally consolidated clay. In order to discuss the numerial results in the following section, an instability of the proposed viscoplastic model is reviewed based on the results by Oka et al.(2002) in a same manner by (Oka et al. 1995). We herein consider the response of the model under conventional triaxial undrained creep conditions using the same method by Oka et al.(1995). Under undrained creep conditions, a constant deviatoric stress is maintained, although the mean effective stress may change due to the undrained condition that requires a total of zero for the volumetric strain rate.

Under the axisymmetric triaxial testing conditions($\sigma'_{11} > \sigma'_{22} = \sigma'_{33}$, $\sigma'_{ij} = 0(i \neq j)$), the deviator stress is expressed by q which is defined as

$$q = \sigma'_{11} - \sigma'_{22} \tag{27}$$

Since the total volumetric strain is zero, the following relation is obtained after integration, i.e.,

$$\varepsilon^{vp}_{kk} = -\frac{\kappa}{(1+e)} \ln \frac{\sigma'_m}{\sigma'_{mc}} \tag{28}$$

where σ'_{me} is the initial value of the mean effective stress.

Herein, the following variables are used under the triaxial conditions:

$$\eta = \sqrt{3/2}\eta^* = q/\sigma'_m, x_{11} = \sqrt{3/2}x^*_{11} \tag{29}$$

For simplicity, we assume that $M_f = M_m = M(M_f = \sqrt{3/2}M^*_f, M_m = \sqrt{3/2}M^*_m)$.

Under undrained triaxial conditions, and disregarding the deviatoric elastic strain rate, viscoplastic axial strain rate $\dot{\varepsilon}^{vp}_{11}$ becomes

$$\dot{\varepsilon}^{vp}_{11} - C\Phi_2 exp[m'(\frac{1}{M}(\eta - x) + \ln\sigma_m/\sigma'_{me} - A_3 y_m)] \tag{30}$$

where A_3 is the volumetric hardening parameter given by $A_3 = \frac{1+e}{\lambda-\kappa}$.

Then, the evolutional equations for the two hardening parameters, x_{11} and y_m, are given by

$$\dot{x}_{11} = B(A - x_{11})\dot{\varepsilon}^{vp}_{11} \tag{31}$$

$$\dot{y}_m = \dot{y}_{m2} = A_3\dot{\varepsilon}_{kk} \tag{32}$$

where $B = \sqrt{3/2}B^*, A = \sqrt{3/2}A^* = M_f$, and $y_{m1} = 0$.

Let us calculate a rate of strain rate denoted by $\ddot{\varepsilon}^{vp}_{11}$. By examining the sign of the rate of strain rate $\ddot{\varepsilon}^{vp}_{11}$, we can estimate the stability of the material system. If the rate of strain rate is positive, the material undergoes a creep failure.

Upon time differentiation of a viscoplastic strain rate without a second material function, the rate of strain rate is obtained as

$$\ddot{\varepsilon}_{11} = -m'a(\dot{\varepsilon}^{vp}_{11})^2 \tag{33}$$

$$a = [\frac{1+e}{M\kappa}(\eta - M)(\eta - \frac{\lambda}{\lambda - \kappa}M) + \frac{B}{M}(A - x_{11})] \tag{34}$$

In the above derivation, the stress dilatancy relation

$$\frac{\dot{\varepsilon}^{vp}_{kk}}{\dot{\varepsilon}^{vp}_{11}} = M - \eta \tag{35}$$

and the undrained condition

$$\dot{\varepsilon}^{vp}_{kk} = -\frac{\kappa}{1+e}\frac{\dot{\sigma}'_m}{\sigma'_m} \tag{36}$$

are introduced.

We will discuss the instability of the model using Equations (37) and (38) and the assumption that the last term is small, since $A - x_{11} \approx 0$ near the failure state. Firstly, we consider the case of undrained creep for normally consolidated clay in which $\eta < M$. In this case, the following conditions prevail, namely, $M - \eta > 0$, $(\frac{\lambda}{\lambda-\kappa}M - \eta) > 0$, and $A - x_{11} > 0$. Since $a > 0$, the rate of strain rate is negative, which leads to the conclusion that the material system is structurally stable in terms of Liapunov.

Next, we will consider the second material function for normally consolidated clay. The second material function is simplified with the assumption that the initial value of $x_{11(0)}$ is zero; i.e., $x_{11} = 0$. Thus,

$$\Phi_2 = 1 + \frac{M\eta}{G_2(M - \eta)} = \frac{G_2(M - \eta) + M\eta}{G_2(M - \eta)} \tag{37}$$

in which $G_2 = \sqrt{2/3}G_2^*$.

$$\ddot{\varepsilon}_{11} = -m'a(\dot{\varepsilon}_{11}^{vp})^2 \tag{38}$$

$$a = -\frac{(1+e)M^2\eta}{m'\kappa(G_2(M-\eta)+M\eta)} + \frac{1+e}{M\kappa}(\eta-M)(\eta-\frac{\lambda}{\lambda-\kappa}M) + \frac{B}{M}(\Lambda_3-x) \tag{39}$$

In contrast to the case without a second material function, it is found that when a second material function is included, the rate of strain rate $\ddot{\varepsilon}_{11}^{vp}$ may become positive before η reaches M, since the first term in the square brackets of Equation (39) increases with a negative sign. Hence, the introduction of a second material function is inevitable for describing the creep failure of normally consolidated clay in the case of monotonically increasing hardening function (see Figure 3). This is consistent with previous results of tests on the instability of normally consolidated clay obtained by Adachi et al., (1990), Oka et al. (1995).

Then, we will discuss the stability of the model in the region where $\eta > M$. This condition corresponds to the undrained creep tests which can be achieved by applying deviator stress q with a small initial mean effective stress located in the overconsolidated region. When there is no second material function, the term a becomes positive in region $M < \eta < \frac{\lambda}{\lambda-\kappa}M$, as shown in Figure 4. In this region, the material becomes unstable due to the fact that $\ddot{\varepsilon}_{11}^{vp} > 0$. On the other hand, when second material function Φ_2 is introduced, the stability cannot be evaluated with Equation (39) in the region where $\eta > M$ because $\xi(=(M\eta)/[G_2(M-\eta)])$ becomes negative due to a simplification which involves the replacement of η_{ij}^{**}, defined by the stress history tensor, with η_{ij}^* based on the stress tensor.

The above considerations show that it is evident that the model can simulate the instability associated with "undrained creep failure" in the specific stress regions. It is worth noting that in the region where $\eta > M$, the model can be unstable even if a second material function is not included. Using an isotropic hardening viscoplastic model for normally consolidated clay, Oka et al.(1994) found that the material model developed by Adachi and Oka (1982) is always stable without a second material function, i.e., $\Phi_2 = 1$. in the normally consolidated region. From the above consideration, however, it becomes evident that overconsolidated clay becomes unstable in region($\eta > M$) even for models without a second material function.

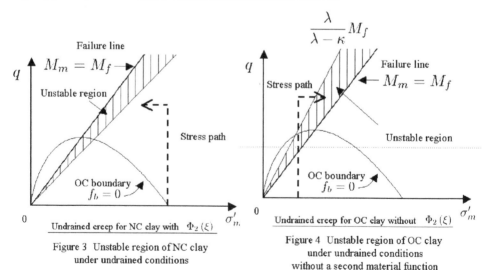

Figure 3 Unstable region of NC clay under undrained conditions

Figure 4 Unstable region of OC clay under undrained conditions without a second material function

4 FINITE ELEMENT ANALYSIS OF STRAIN LOCALIZATION BY AN ELASTO-VISCO-PLASTIC MODEL

Oka et al. (1994, 1995, 2002) numerically studied the strain localization problem using an elasto-viscoplastic model for normally consolidated clay based on a Perzyna type of overstress model and a Cam-clay model. In this study, the extended model which can simulate the behavior of overconsolidated clay as well as normally consolidated clay is used in numerical analysis in order to study the effects of dilatancy and strain rate on strain localization.

Numerical solutions for the plane strain compression problems of water-saturated clay are obtained via the finite element method. In the finite element analysis, the updated Lagrangian method with the objective Jaumann rate of Cauchy stress is used for a weak form of the equilibrium equation. For describing the motion of pore water, a Biot type of two-phase mixture theory is used in the analysis with a $v_i(velocity) - u_w(pore\ pressure)$ formulation (see Oka et al. 2000).

In the finite-element formulation, the tangent modulus method (Pierce et al. 1984) is used. An eight-node quadrilateral element with a reduced Gaussian (2×2) integration is used to eliminate the shear locking and to reduce the appearance of a spurious hourglass mode. On the other hand, the pore water pressure is defined at four corner nodes. A weak form of the continuity equation is integrated with a (2×2) full integration. Using this combination of the spatial integration scheme, the effective stresses, the pore water pressure, and the strain are all calculated at the same integration points for each element.

Figure 5 shows the size of the specimen and the associated boundary conditions. As a trigger for strain localization, horizontal displacements on both top and bottom surface edges are constrained. Relaxation of this constraint through the introduction of a frictional boundary will be discussed later. The material parameters used in the analysis are listed in Table 1 with the coefficient of permeability and the K_0 values. In the analysis, the time increment is determined by the increment of average strain $\Delta \varepsilon_{11} = 0.01\%$. In this case, a time increment of 6 sec is used.

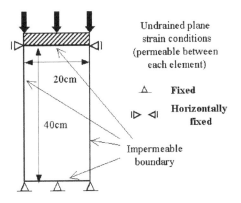

Figure 5 Size of specimen and boundary conditions

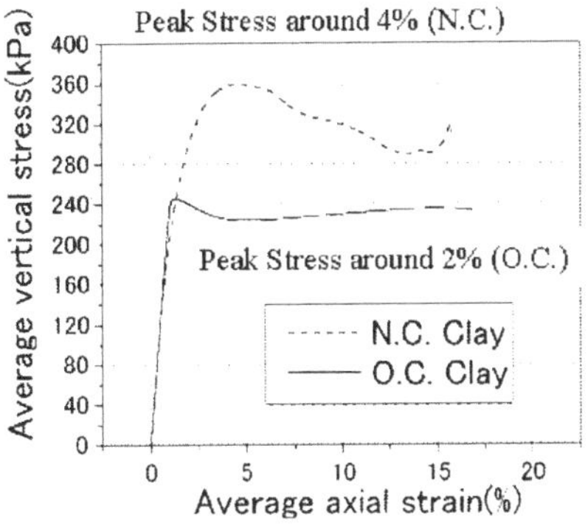

Figure 6 Stress-strain relations for NC and OC clays
(0.1%/min, k = 1.54 × 10⁻⁸m/s)

4.1 Effects of dilatancy

The compression of a clay specimen is simulated under globally undrained plane strain conditions. Compression is performed under displacement control with average strain rates of $0.1\%/\text{min}$ and $1\%/\text{min}$. Figure 6 shows the average stress-strain relationships and Figure 7 shows the simulated results for normally consolidated and overconsolidated clays with a permeability coefficient of $1.54 \times 10^{-8}(\text{m/s})$.

It can be seen from Figure 7 that the deformed meshes of normally and overconsolidated clay specimens display localization of deformation at an average axial strain of 8% and 6%, respectively. The appearance of the shear band at a larger strain in N.C. clay is consistent with the stress-strain curves with gradual softening. The occurrence of localization at an early stage of deformation in the case of overconsolidated clay is consistent with the average stress-strain relationships shown in Figure 6. This tendency has been observed in the experiments (e.g., Hicher et al. 1994).

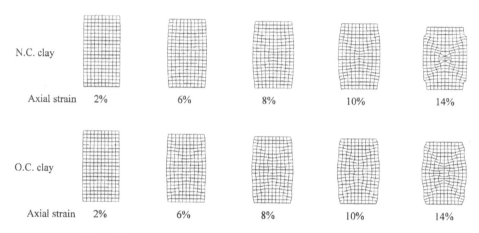

Figure 7 Deformed meshes of NC and OC clays (0.1%/min, 1.54 × 10⁻⁸(m/s))

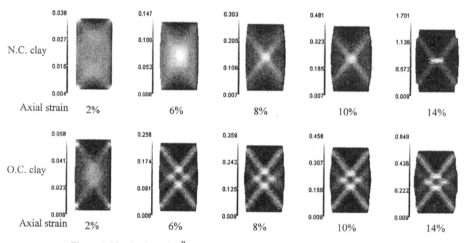

Figure 8 Distribution of γ^p for NC and OC clays (0.1%/min, 1.54 × 10⁻⁸(m/s))

Figure 8 shows the accumulated viscoplastic shear strain $\gamma^p \equiv \int \sqrt{de_{ij}^{vp} de_{ij}^{vp}}$. In the case of overconsolidated clay, strain localization starts near the edges of the top and the bottom plates, and finally, four shear bands appear. In contrast to the case of overconsolidated clay, only two shear bands are seen for normally consolidated clay, with shear bands clearly developing just beneath the edges of the top and the bottom plates. As for the distribution of viscoplastic volumetric strain magnitude, it is seen from Figure 9(a) that a decrease in viscoplastic volumetric strain (viscoplastic volume expansion) occurs along the shear bands for overconsolidated clay, while only viscoplastic compression is seen in the case of normally consolidated clay. The tendency of the distribution of viscoplastic volumetric strain is, in fact, related to the changes in mean effective stresses since calculations are carried out under globally undrained conditions. Figure 9(b) shows the distribution of mean effective stress at an axial strain of 10%. In the case of overconsolidated clay, the mean effective stress increases in the specimen from its initial value, i.e., $100\ kPa$. However, along the shear bands, mean effective stress levels are lower than those in the other regions of the specimen. In the case of normally consolidated clay, the mean effective stress decreases from its initial value of $392\ kPa$ due to negative dilatancy. The extent of the decrease in mean effective stress is larger in shear bands.

In general the distribution of mean effective stress is related to pore fluid motion. Hence, in order to examine the distribution of mean effective stress, it is necessary to evaluate the effect of the permeability coefficient. Figure 10 shows the distributions of mean effective stress and viscoplastic volumetric strain for overconsolidated clay with a permeability coefficient as low as 1.54×10^{-10}(m/s) at an axial strain of 10%. In this case, the mean effective stress along the shear bands is relatively higher than that in the regions between them. This tendency is in contrast to that found in the numerical results obtained for the high permeability case shown in Figure 9(b). The reason for this difference is that pore water can easily move within a material with high permeability. Hence, an increase in mean effective stress due to positive dilatancy can be cancelled by the inflow of pore water toward the shear bands. Figure 11 displays the distribution of pore water pressure. Comparing the distributions of mean effective stress and plastic shear strain, it is seen that the distribution of pore water pressure is rather homogeneous for both normally and overconsolidated clays. However, higher levels of pore water pressure develop for normally consolidated clay. The relatively homogeneous distribution of pore water pressure within the specimen is considered to be due to the migration of pore water.

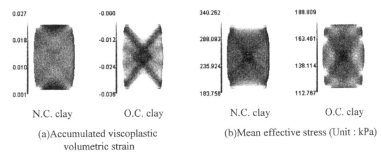

| N.C. clay | O.C. clay | N.C. clay | O.C. clay |

(a)Accumulated viscoplastic volumetric strain

(b)Mean effective stress (Unit : kPa)

Figure 9 Distribution of accumulated viscoplastic volumetric strain and mean effective stress for NC and OC clay (axial strain : 10%, strain rate : 0.1%/min, k = 1.54 × 10⁻⁸m/s)

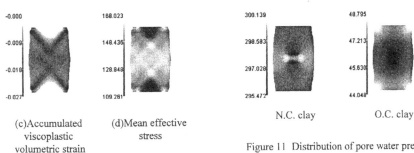

(c)Accumulated viscoplastic volumetric strain

(d)Mean effective stress

Figure 10 Distribution of accumulated volumetric viscoplastic strain and mean effective stress for OC clay (axial strain : 10%, strain rate : 0.1%/min, k = 1.54 × 10⁻¹⁰m/s)

N.C. clay   O.C. clay

Figure 11 Distribution of pore water pressure for NC and OC clays (axial strain : 10%, strain rate : 0.1%/min, k = 1.54 × 10⁻⁸m/s)

4.2 Effects of strain rate

Stress-strain relations and distribution of γ^p for different average strain rates (0.1%/min and 1%/min) are shown in Figure 12; the strain rate sensitivities are clearly seen in the stress-strain curves. It is seen that normally consolidated clay is more stable with higher strain rates, while the effect of strain rates is not very significant for overconsolidated clay. The reason for this observation is that, for normally consolidated clay, the mean effective stress greatly changes before the peak stress owing to the shape of the plastic potential function. This is not the case for overconsolidated clay. In Figure 12, the appearance of four shear bands is clearly seen for the cases of higher strain rate. In addition, the figure shows that the distance between the shear bands is wider than that in the case of lower strain rate. A possible reason for the larger number of shear bands in the overconsolidated clay than in the normally consolidated clay is that the strain softening associated with dilation may cause a reduction in the mean effective stress. Consequently, the adjacent regions can easily deform so as to lead to the occurrence of a larger number of shear bands.

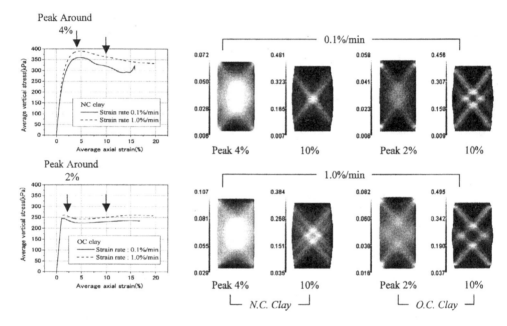

Figure 12 Stress-strain relations and distribution of γ^p at peak stress and axial strain of 10% (k = 1.54×10⁻⁸m/s)

4.3 Mesh size dependency

In order to study the mesh size dependency of the numerical results, simulations are performed using various mesh sizes. For normally consolidated clay, there is no significant mesh size dependency as shown in Figure 13. As for the overconsolidated clay, numerical calculations with smaller elements (800 elements) diverged after the peak stress around 2% of the axial strain. In order to examine the effect of the end constraints, the "no lateral displacements at top and bottom plates" was relaxed by instead using a frictional boundary with a frictional coefficient of 0.00001. Figure 14 shows the deformed meshes and distribution of γ^p for different numbers of elements. Since no noticeable mesh size dependency is observed in Figure 14, the occurrence of any numerical instability can only result from the imposition of strong constraints. No influence of mesh size on stress-strain responses is seen in Figure 15. Hence, it is worth noting that the numerical calculations for overconsolidated clay are more sensitive to constraint conditions than those for normally consolidated clay. In other words, O.C. clay easily leads to instability. This tendency is consistent with the results of the instability analysis obtained in Chapter 3.

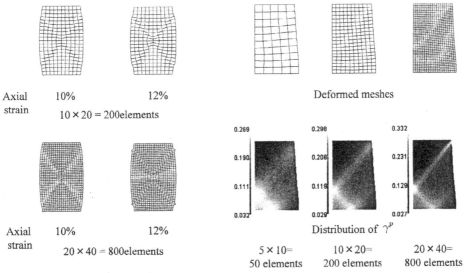

Figure13 Deformed meshes of NC clay
for 200 and 800 elements
(0.1%/min, 1.54 × 10⁻⁸m/s)

Figure 14 Deformation and distribution of γ^P for O.C. clay
(axial strain : 10%, strain rate : 0.1%/min, 1.54 × 10⁻⁸m/s,
coefficient of friction : 0.00001)

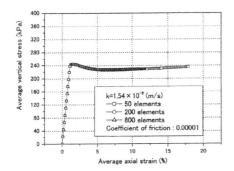

Figure 15 Average stress-strain relations of OC clay
with different element sizes
(0.1%/min, 1.54 × 10⁻⁸m/s)

5 CONCLUSIONS

The obtained conclusions from the presnt study are as follows. An elasto-viscoplastic constitutive model for clay based on a Chaboche type of viscoplasticity theory was used in the analysis. The model can describe both positive and negative dilatancy characteristics which are the important characteristics of soil. The instability of the model was revisited under undrained triaxial creep conditions. It was seen that the model with positive dilatancy was more unstable than the model with negative dilatancy in terms of creep failure. Even when a second material function was not included in the formulation, the model could become unstable with positive dilatancy. On the other hand, the model with negative dilatancy became unstable only when a second material function was introduced. As for the numerical simulation of the shear band development, using the elasto-viscoplastic model, it was found that the strain localization pattern was strongly affected by ditatancy characteristics as well as the strain rate.

REFERENCES

Adachi, T. and Oka, F., (1982). Constitutive equations for normally consolidated clay based on elasto-viscoplasticity, Soils and Foundations, 22, 4, 57-70.

Adachi, T., Oka, F. and Mimura, M., (1987). An elasto-viscoplastic theory for clay failure, Proc. 8th Asian Regional Conference on Soil Mechanics and Foundation Engineering, Kyoto, 1, Japanese Society for Soil Mechanics and Foundation Engineering, 5-8.

Adachi, T., Oka, F. and Mimura, M., (1990). Elasto-viscoplastic constitutive equations and their application to consolidation analysis, J. of Engineering Materials and Technology, ASME, 112, 202-209.

Adachi, T. and Oka, F., 1995. An elasto-plastic constitutive model for soft rock with strain softening, Int. J. Numerical and Analytical Methods in Geomechanics, 19, 233-247.

Aifantis, E.C., (1984). On the microstructural origin of certain inelastic models, ASME, J. Engineering Materials and Tech., 106, 326-330.

Aifantis, E.C., (1987). The physics of plastic deformation, International Journal of Plasticity, 3, 211- 247.

Chaboche, J.L. and Rousselier, G., (1980). On the plastic and viscoplastic constitutive equations-Part I: Rules developed with internal variable concept, J. Pressure Vessel tech., ASME, 105, 103-158.

Hicher, P.Y., Wahyudi, H. and D. Tessier, (1994). Microstructural analysis of strain localisation in clay, Computers and Geotechnics, 16, 205-222.

Loret, B and Prevost, J.H., (1991). Dynamic strain localization in fluid-saturated porous media, J. Engineering Mechanics, ASCE, 117,4,907-922.

Mühlhaus, H.-B. and Aifantis, E.C., (1991). A variational principle for gradient plasticity, Int. J. of Solids and Structures, 28, 7, 845-857.

Oka, F., (1985). Elasto/viscoplastic constitutive equations with memory and internal variables, Computer and Geotechnics, 1, 59-69.

Oka, F., Adachi, T. and Yashima, A., (1994). Instability of an elasto-viscoplastic constitutive model for clay and strain localization, Mechanics of Materials, 18, 119-129.

Oka, F., Adachi, T. and Yashima, A., (1995). A strain localization analysis of clay using a strain softening viscoplastic model, Int. Journal of Plasticity, 11, 5, 523-545.

Oka, F., A. Yashima, K. Sawada and E.C. Aifantis, (2000). Instability of gradient-dependent elastoviscoplastic model for clay and strain localization, Computer Methods in Applied Mechanics and Engineering, 183, 67-86.

Oka, F., Higo, Y. and Mingjing Jiang, (2000). Effects of material inhomogeneity and transport of pore water on strain localization analysis of fluid-saturated strain gradient viscoplastic geomaterial, Proc. of 8th International Symposium on Plasticity and Current Applications, Whistler, A.S. Khan, H. Zhang and Y. Yuan eds., Neat Press, 306-308.

Oka, F., Higo, Y. and Kimoto, S., (2002). Effect of dilatancy on the strain localization of water-saturated elasto-viscoplastic soil, Int. J. of Solids and Structures, Vol.39, Issues 13-14, pp.3625-3647.

Pierce, D., Shih, C.F. and Needleman, A., (1984). A tangent modulus method for rate dependent solids, Computers and Structures, 18, 5, 875-887.

Perzyna, P., (1963). The constitutive equations for work-hardening and rate sensitive plastic materials, Proc. Vibrational Problems, Warsaw, 4, 3, 281-290.

Rice, J., (1975). On the stability of dilatant hardening for saturated rock masses, J. Geophys. Res., 80,11,1531-1536.

Rice, J., (1976). The localization of plastic deformation, IUTAM symposium, Theoretical and Applied Mechanics, eds. Koiter, W.T., Noth-Holland, 207-220.

Schrefler, B.A., Sanavia, L. and Majorana, C.E., (1996). A multiphase medium model for localisation and postlocalisation simulation in geomaterials, Mechanics of Cohesive-Fictional Materials, 1,1 ,95-114.

Vardoulakis, I. and Aifantis, E.C., 1991. A gradient flow theory of plasticity for granular materials, Acta Mechanica 87, 197-217.

Bifurcations & Instabilities in Geomechanics, – Labuz & Drescher (eds.)
© *2003 Swets & Zeitlinger, Lisse, ISBN 90 5809 563 0*

Adaptive computation of localization phenomena in geotechnical applications

W. Ehlers & T. Graf
Institute of Applied Mechanics (CE), University of Stuttgart, Germany

ABSTRACT: The numerical treatment of unsaturated soil plays a dominant role in geotechnical engineering, as for instance in the investigation of deformation and localization behaviour of the subsurface. Proceeding from the second law of thermodynamics, unsaturated soil is described within the well-founded framework of the *Theory of Porous Media* (TPM), by a porous, materially incompressible, elasto-plastic or elasto-viscoplastic solid skeleton (soil), a viscous, materially incompressible pore-liquid (water) and a viscous, materially compressible pore-gas (air). Based on quasi-static considerations, numerical computations are based on weak formulations of the momentum balance of the overall triphasic medium together with the mass balance equations of the pore-fluids. The resulting system of strongly coupled differential-algebraic equations can be solved by use of the finite element tool *PANDAS*. Finally, the formulation is applied to various initial-boundary-value problems, as e. g. the simulation of the fluid-flow situation or the slope failure of embankments.

1 INTRODUCTION

For the description of ground water-flow situations or stability examinations of embankments, an overall description of partially saturated soil is required. In the present contribution, frictional materials are considered within the well-founded framework of the *Theory of Porous Media* (TPM) (Bowen 1980, Ehlers 1993, Ehlers 2002) defined as the mixture theory (Truesdell & Toupin 1960, Bowen 1976) extended by the concept of volume fractions. In this paper, partially saturated soil is described by a model consisting of a materially incompressible solid skeleton, a materially incompressible pore-liquid and a materially compressible pore-gas.

The paper is organized as follows: In Section 2, a triphasic model for partially saturated soil is presented. Therein, the porous solid skeleton is described by a general elasto-viscoplastic approach within the geometrically linear case. The interactions between the solid skeleton and the viscous pore-fluids are taken into account by the capillary-pressure-saturation relation and *Darcy*-like filter laws including relative permeabilities to consider the interactions between the pore-fluid mobilities.

In Section 3, the numerical treatment of the strongly coupled solid-fluid problem is described on the basis of the Finite Element Method (FEM) and illustrated in Section 4 by examples of gravity driven initial-boundary-value problems, solved by the finite element tool *PANDAS* (*P*orous media *A*daptive *N*onlinear finite element solver based on *D*ifferential *A*lgebraic *S*ystems) (Ehlers & Ellsiepen 1998).

2   A TRIPHASIC MODEL FOR PARTIALLY SATURATED SOIL

2.1   *The concept of volume fractions*

Based on the fundamental concepts of the *Theory of Porous Media*, one proceeds from the assumption that all individual materials are assumed to be in a state of ideal disarrangement, i. e. they are statistically distributed over the control space. This assumption together with the prescription of a real or a virtual averaging process leads to a model φ of superimposed and interacting continua φ^α,

$$\varphi = \sum_\alpha \varphi^\alpha,$$
(1)

where the index $\alpha = S$ denotes the solid constituent, $\alpha = L$ the liquid constituent and $\alpha = G$ the gas constituent. The mathematical functions for the description of the geometrical and physical properties of the individual materials are field functions defined all over the control space. The volume V of the overall multiphasic aggregate $\mathcal{B}$ results from the sum of the partial volumes of the constituents φ^α in $\mathcal{B}$,

$$V = \int_\mathcal{B} dv = \sum_\alpha V^\alpha,$$
(2)

where

$$V^\alpha = \int_\mathcal{B} dv^\alpha =: \int_\mathcal{B} n^\alpha \, dv.$$
(3)

Following this, the volume fraction n^α is defined as the ratio of the volume element dv^α of a given constituent φ^α with respect to the volume element dv of the overall medium φ at a local point:

$$n^\alpha := \frac{dv^\alpha}{dv}.$$
(4)

Since, in general, there is no void space in the overall medium, the saturation condition

$$n^S + n^F = 1, \quad n^F = n^L + n^G$$
(5)

holds, where n^F defines the total fluid volume fraction or the porosity, respectively. With respect to the porosity, the saturation functions are defined as

$$s^\beta := \frac{n^\beta}{n^F}, \quad \beta \in \{L, G\}$$
(6)

with the corresponding constraint

$$s^L + s^G = 1.$$
(7)

By use of the definition (4), two different density functions of each constituent φ^α can be introduced:

$$\rho^{\alpha R} = \frac{dm^\alpha}{dv^\alpha}, \quad \rho^\alpha = \frac{dm^\alpha}{dv}.$$
(8)

Therein, the material (realistic or effective) density $\rho^{\alpha R}$ relates the local mass dm^α to the volume element dv^α, whereas the partial (global or bulk) density ρ^α relates the same mass to the volume element dv. Following this, the density functions are related to each other by

$$\rho^\alpha = n^\alpha \rho^{\alpha R}.$$
(9)

The density ρ of the overall medium (mixture density) is then given by the sum over all partial densities ρ^α:

$$\rho = \sum_\alpha \rho^\alpha.$$
(10)

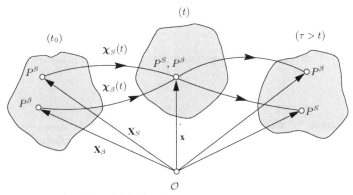

Figure 1: Motion of a multiphasic mixture.

2.2 Kinematical relations

Proceeding from mixture theories as the fundamental basis of the *Theory of Porous Media*, one directly makes use of the concept of superimposed continua with internal interactions and individual states of motion. In the framework of this concept, each spatial point **x** of the current configuration is, at any time t, simultaneously occupied by material particles (material points) P^α of all constituents φ^α. These particles proceed from different reference positions at time t_0, cf. Figure 1. Thus, each constituent is assigned its own motion function

$$\mathbf{x} = \boldsymbol{\chi}_\alpha(\mathbf{X}_\alpha, t). \tag{11}$$

As a result, each spatial point **x** can only be occupied by one single material point P^α of each constituent φ^α. The assumption of unique motion functions, where each material point P^α of the current configuration has a unique reference position $\mathbf{X}_\alpha$ at time t_0, requires the existence of unique inverse motion functions $\boldsymbol{\chi}_\alpha^{-1}$ based on non-singular *Jacobian* determinants J_α:

$$\mathbf{X}_\alpha = \boldsymbol{\chi}_\alpha^{-1}(\mathbf{x}, t), \quad J_\alpha = \det\frac{\partial\boldsymbol{\chi}_\alpha}{\partial\mathbf{X}_\alpha} \neq 0. \tag{12}$$

It follows from (11) that each constituent has its own velocity and acceleration fields. In the basic *Lagrange*an setting, these fields are given by

$$\overset{\prime}{\mathbf{x}}_\alpha = \frac{\partial\boldsymbol{\chi}_\alpha(\mathbf{X}_\alpha, t)}{\partial t}, \quad \overset{\prime\prime}{\mathbf{x}}_\alpha = \frac{\partial^2\boldsymbol{\chi}_\alpha(\mathbf{X}_\alpha, t)}{\partial t^2}. \tag{13}$$

With the aid of the inverse motion function $(12)_1$, an alternative formulation of (13) leads to the *Euler*ian description

$$\overset{\prime}{\mathbf{x}}_\alpha = \overset{\prime}{\mathbf{x}}_\alpha(\mathbf{x}, t), \quad \overset{\prime\prime}{\mathbf{x}}_\alpha = \overset{\prime\prime}{\mathbf{x}}_\alpha(\mathbf{x}, t). \tag{14}$$

Suppose that Γ is an arbitrary, steady and sufficiently often steadily differentiable scalar function of $(\mathbf{x}, t)$. Then, the material time derivative of Γ following the motion of φ^α reads

$$(\Gamma)'_\alpha = \frac{\mathrm{d}_\alpha}{\mathrm{d}t}\Gamma = \frac{\partial\Gamma}{\partial t} + \operatorname{grad}\Gamma\cdot\overset{\prime}{\mathbf{x}}_\alpha. \tag{15}$$

Therein, the operator "grad $(\cdot)$" denotes the partial derivative of $(\cdot)$ with respect to the local position **x**.

Describing coupled solid-fluid problems, it is convenient to proceed from a *Lagrange*an description of the solid matrix using the solid displacement vector $\mathbf{u}_S$ as the primary kinematic variable, whereas the pore-fluids are favourably described in a modified *Euler*ian setting by use of the seepage velocities $\mathbf{w}_\beta$ describing the fluid motions with respect to the deforming skeleton material:

$$\mathbf{u}_S = \mathbf{x} - \mathbf{X}_S, \quad \mathbf{w}_\beta = \overset{\prime}{\mathbf{x}}_\beta - \overset{\prime}{\mathbf{x}}_S. \tag{16}$$

From (11) and (12)$_1$, one obtains the material deformation gradient $\mathbf{F}_\alpha$ and its inverse $\mathbf{F}_\alpha^{-1}$ by

$$\mathbf{F}_\alpha = \text{Grad}_\alpha \, \mathbf{x} \,, \qquad \mathbf{F}_\alpha^{-1} = \text{grad} \, \mathbf{X}_\alpha \,. \tag{17}$$

Therein, the operator "$\text{Grad}_\alpha(\,\cdot\,)$" denotes the partial derivative of $(\,\cdot\,)$ with respect to the reference position $\mathbf{X}_\alpha$ of φ^α.

2.3 Balance relations

The discussion of balance relations for multiphasic materials is based on *Truesdell*'s "metaphysical principles" of mixture theories (Truesdell 1984). The foundation of *Truesdell*'s principles proceeds from the idea that both the balance relations of the constituents φ^α and the balance relations of the overall medium φ can be given in analogy to the balance relations of classical continuum mechanics of single-phase materials, extended by so-called production terms to describe the interaction mechanisms between the constituents.

Following this, the partial balance equations of mass

$$(\rho^\alpha)'_\alpha + \rho^\alpha \, \text{div} \, \mathbf{\dot{x}}_\alpha = 0 \tag{18}$$

and of linear momentum

$$\rho^\alpha \mathbf{\ddot{x}}_\alpha = \text{div} \, \mathbf{T}^\alpha + \rho^\alpha \mathbf{g} + \mathbf{\hat{p}}^\alpha \tag{19}$$

hold (Ehlers 2002). Therein, "$\text{div}(\cdot)$" is the divergence operator corresponding to "$\text{grad}(\cdot)$", $\mathbf{g}$ is the overall gravity, $\hat{\rho}^\alpha$ is the mass production and $\mathbf{\hat{p}}^\alpha$ is the direct linear momentum production. To conserve the linear momentum of the whole mixture, the total linear momentum production must be zero yielding

$$\mathbf{\hat{p}}^S + \mathbf{\hat{p}}^F = \mathbf{0}, \qquad \mathbf{\hat{p}}^F = \mathbf{\hat{p}}^L + \mathbf{\hat{p}}^G \,. \tag{20}$$

If φ^α is a materially incompressible constituent ($\rho^{\alpha R} = \text{const.}$), the mass balance equation (18) reduces to the volume balance

$$(n^\alpha)'_\alpha + n^\alpha \, \text{div} \, \mathbf{\dot{x}}_\alpha = 0 \,. \tag{21}$$

By integration of (21), the volume fraction n^S of the solid skeleton is given by

$$n^S = n^S_{0S}(\det \mathbf{F}_S)^{-1} \,, \tag{22}$$

where a formal linearization around the natural state of φ^S leads to the geometrically linear version

$$n^S = n^S_{0S}(1 - \text{div} \, \mathbf{u}_S) \,. \tag{23}$$

Therein, n^S_{0S} is the volume fraction of the solid skeleton in its reference configuration.

To describe the fluid constituents with respect to the moving solid phase, the partial mass balances of the pore-fluids must be reformulated. Therefore, the material time derivative following the fluid motion must be transformed to a time derivative with respect to the solid motion and a modified convective term:

$$(\rho^\beta)'_\beta = (\rho^\beta)'_S + \text{grad} \, \rho^\beta \cdot (\mathbf{\dot{x}}_\beta - \mathbf{\dot{x}}_S) \,. \tag{24}$$

Inserting (24) into (18) yields the balance of mass for the fluid constituents with respect to the motion of the solid skeleton:

$$(\rho^\beta)'_S + \rho^\beta \, \text{div} \, \mathbf{\dot{x}}_S + \text{div}(\rho^\beta \mathbf{w}_\beta) = 0 \,. \tag{25}$$

The formulation of initial-boundary-value problems of partially saturated soil based on a triphasic model includes five primary variables, the kinematic variables $\mathbf{u}_S$, $\mathbf{w}_L$, $\mathbf{w}_G$ and the effective liquid and gas pressures p^{LR} and p^{GR}. For geotechnical problems, the resulting system of equations consists of the linear momentum balances of all constituents under quasi-static conditions ($\mathbf{\ddot{x}}_\alpha = 0$), the volume balance of the pore-liquid and the mass balance of the pore-gas.

2.4 Constitutive equations

To close the model under consideration, constitutive equations are required for the partial *Cauchy* stress tensors $\mathbf{T}^\alpha$, the linear momentum production terms $\hat{\mathbf{p}}^\beta$ of the pore-fluids, the effective pressure p^{GR} of the pore-gas and the saturation s^L of the pore-liquid.

An evaluation of the entropy inequality leads to the fact that the solid and the fluid stresses $\mathbf{T}^\alpha$ as well as the linear momentum production $\mathbf{p}^\beta$ consist of two parts (Bowen 1980, Ehlers 1993). The first one is governed by the pore-pressure, while the second one, the so-called "extra term", results from the solid deformations (effective stress) or the pore-fluid flow (frictional stress):

$$\mathbf{T}^S = -n^S p \mathbf{I} + \mathbf{T}^S_E,$$
$$\mathbf{T}^\beta = -n^\beta p^{\beta R} \mathbf{I} + \mathbf{T}^\beta_E, \tag{26}$$
$$\hat{\mathbf{p}}^\beta = p^{\beta R} \operatorname{grad} n^\beta + \hat{\mathbf{p}}^\beta_E.$$

Therein, the pore-pressure p is given in analogy to *Dalton*'s law by

$$p = s^L p^{LR} + s^G p^{GR}. \tag{27}$$

Note that for the materially incompressible pore-liquid, the pressure p^{LR} acts as a *Lagrange*an multiplier and is simply determined by the boundary conditions of the problem under study.

In the remainder of this article, the pressures p^{LR} and p^{GR} are understood as the *effective excess pressures* exceeding a typical surrounding pressure like, e. g., the atmospheric pressure p_0. Furthermore, it should be mentioned that *Dalton*'s law is more than only a convenient expression, since it can be recovered from thermodynamical considerations to satisfy the entropy principle.

In geotechnical problems, the fluid friction forces $\operatorname{div} \mathbf{T}^\beta_E$ can be neglected in comparison to the viscous interaction terms $\hat{\mathbf{p}}^\beta_E$. Thus, the overall *Cauchy* stress $\mathbf{T}$ given by the sum of all partial stress tensors of the constituents yields the well-known *concept of effective stress* (Bishop 1959):

$$\mathbf{T} = -p\mathbf{I} + \mathbf{T}^S_E. \tag{28}$$

2.4.1 The fluid constituents

The "extra term" of the linear momentum production between the pore-fluids $(26)_3$ is related to the seepage velocity $\mathbf{w}_\beta$ by

$$\hat{\mathbf{p}}^\beta_E = -(n^\beta)^2 \gamma^{\beta R} (\mathbf{K}^\beta_r)^{-1} \mathbf{w}_\beta, \tag{29}$$

where $\gamma^{\beta R}$ is the specific weight and $\mathbf{K}^\beta_r$ the *relative* permeability tensor of the constituent φ^β related to the *Darcy* permeability tensor $\mathbf{K}^\beta$ via

$$\mathbf{K}^\beta_r = \kappa^\beta_r \mathbf{K}^\beta. \tag{30}$$

Therein, $\mathbf{K}^\beta$ is understood as the permeability tensor of φ^β specified for fully saturated conditions $(s^\beta = 1)$ and depending on the actual deformation of the solid skeleton, whereas κ^β_r is the relative permeability depending on the saturation of φ^β. $\mathbf{K}^\beta$ depends on the intrinsic permeability $\mathbf{K}^S$ through

$$\mathbf{K}^\beta = \frac{\gamma^{\beta R}}{\mu^{\beta R}} \mathbf{K}^S. \tag{31}$$

Note that this equation can be used to relate the *Darcy* permeability tensor $\mathbf{K}^\beta$ of various fluids through their specific weight $\gamma^{\beta R}$ and their effective shear viscosity $\mu^{\beta R}$ to the intrinsic permeability $\mathbf{K}^S$. In order to describe the deformation dependence of the intrinsic permeability, it is assumed (Eipper 1998) that

$$\mathbf{K}^S = \left(\frac{1-n^S}{1-n^S_{0S}}\right)^\pi (\mathbf{B}_S)^\xi \mathbf{K}^S_{0S}, \tag{32}$$

251

where $\overset{+}{\mathbf{B}}_S = \frac{1}{2}(\mathbf{B}_S \divideontimes \mathbf{B}_S)$ is the adjoint or the cofactor of the left *Cauchy-Green* tensor $\mathbf{B}_S = \mathbf{F}_S \mathbf{F}_S^T$ or the Finger tensor, respectively. $\mathbf{K}_{0S}^S$ is the intrinsic permeability tensor of the undeformed skeleton, whereas π and ξ are material parameters. In particular, in case of an initially isotropic solid, $\mathbf{K}_{0S}^S$ reduces to

$$\mathbf{K}_{0S}^S = K_{0S}^S \mathbf{I}, \tag{33}$$

where K_{0S}^S is the initial intrinsic permeability coefficient related to the initial *Darcy* permeability coefficient k_{0S}^β in analogy to (31):

$$k_{0S}^\beta = \frac{\gamma^{\beta R}}{\mu^{\beta R}} K_{0S}^S. \tag{34}$$

It was shown by the first author and coworkers (Ehlers et al. 1999, Eipper 1998) that ξ governs the deformation induced part of the anisotropic permeability. Thus, if the material is fully isotropic, ξ reduces to zero and (32) yields

$$\mathbf{K}^S = \left(\frac{1-n^S}{1-n_{0S}^S} \right)^\pi K_{0S}^S \mathbf{I}. \tag{35}$$

Concerning partially saturated soil, three zones can be separated, cf. Figure 2. Beneath the ground-water table, most of the pore-space is filled with the pore-liquid, which is mobile with the *Darcy* permeability measured under fully saturated conditions. Nevertheless, there is also a small amount of pore-gas trapped with a residual saturation s_{res}^G. A second zone can be found in dependency of the described soil at a certain height over the ground-water level. There, the pore-gas is mobile and the pore-liquid is trapped with a residual saturation s_{res}^L. In the area between these two zones, the partially saturated zone, where both fluids are mobile, a relation between the pore-liquid saturation s^L and the capillary pressure

$$p^C = p^{GR} - p^{LR} \tag{36}$$

defined in analogy to Brooks & Corey (1966) or van Genuchten (1980) can be formulated. Here, this relation is given by a formulation according to van Genuchten (1980) via

$$s_{eff}^L(p^C) = [1 + (\alpha_{gen} p^C)^{j_{gen}}]^{-h_{gen}}, \tag{37}$$

where α_{gen}, j_{gen} and h_{gen} are material parameters. Therein, the effective saturation

$$s_{eff}^L := \frac{s^L - s_{res}^L}{1 - s_{res}^L - s_{res}^G} \tag{38}$$

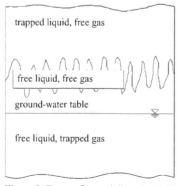
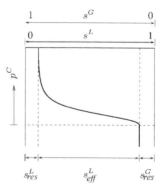

Figure 2: Zones of a partially saturated soil.

252

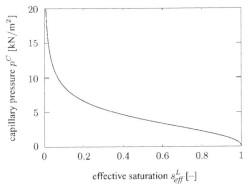

Figure 3: Capillary-pressure-saturation relation according to van Genuchten (1980) with $\alpha_{gen} = 2 \cdot 10^{-4}$, $j_{gen} = 2.3$, $h_{gen} = 1.5$.

describing the area between the two residual saturations is defined following Finsterle (1993), cf. Figure 3. (37) can be extended to describe the hysteretic behaviour of the capillary-pressure-saturation relation which consequently differs for drying and wetting circles because of the specific hydraulic behaviour of the soil, cf. Figure 4.

In the van Genuchten model (van Genuchten 1980), the relative permeability functions are given by

$$
\begin{aligned}
\kappa_r^L &= (s_{\text{eff}}^L)^{\epsilon_{gen}} \left\{ 1 - [1 - (s_{\text{eff}}^L)^{1/h_{gen}}]^{h_{gen}} \right\}^2, \\
\kappa_r^G &= (1 - s_{\text{eff}}^L)^{\gamma_{gen}} [1 - (s_{\text{eff}}^L)^{1/h_{gen}}]^{2 h_{gen}},
\end{aligned}
\tag{39}
$$

where ϵ_{gen} and γ_{gen} are further parameters to describe the hydraulic behaviour of the soil, cf. Figure 5. If the effective saturation is zero, κ_r^L is zero implicating the immobility of the pore-liquid. If the effective saturation is one, κ_r^L is one leading to a pore-liquid which is mobile with the *Darcy* permeability under fully saturated conditions. For the pore-gas, equivalent conditions hold. For more details, the interested reader is referred to the work by Helmig (1997) and Vogler (1999).

Inserting (29) into the quasi-static fluid linear momentum balance leads to the *Darcy* equation

$$
n^\beta \mathbf{w}_\beta = -\frac{\mathbf{K}_r^\beta}{\gamma^{\beta R}} \left(\operatorname{grad} p^{\beta R} - \rho^{\beta R} \mathbf{g} \right).
\tag{40}
$$

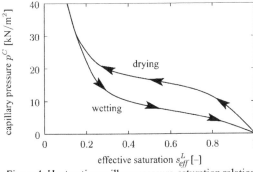

Figure 4: Hysteretic capillary-pressure-saturation relation (qualitative).

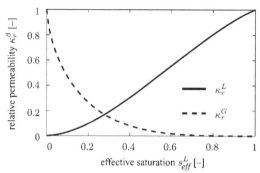

Figure 5: Relative permeability according to van Genuchten (1980) with $\epsilon_{gen} = 0.5$, $\gamma_{gen} = 0.333$.

The effective density function of the materially compressible pore-gas is assumed to be governed by the ideal gas law (*Boyle-Mariotte*'s law):

$$\rho^{GR} = \frac{p^{GR} + p_0}{\bar{R}^G \, \theta} \, . \tag{41}$$

Therein, $\bar{R}^G$ denotes the specific gas constant of the pore-gas and θ the absolute *Kelvin*'s temperature. However, in the present investigations, it is assumed that the overall model can be described under isothermal conditions ($\theta = \text{const.}$).

2.4.2 The solid constituent

Following the geometrically linear approach to elasto-plasticity, the solid strain tensor ε_S is additively decomposed into an elastic and a plastic part:

$$\varepsilon_S = \tfrac{1}{2} [\, \text{grad} \, \mathbf{u} + (\text{grad} \, \mathbf{u})^T \,] = \varepsilon_{Se} + \varepsilon_{Sp} \, . \tag{42}$$

The solid extra stress is governed by the generalized *Hooke*an law

$$\mathbf{T}_E^S = 2 \, \mu^S \, \varepsilon_{Se} + \lambda^S (\, \varepsilon_{Se} \cdot \mathbf{I} \,) \, \mathbf{I}, \tag{43}$$

where μ^S and λ^S are the *Lamé* constants of the porous material.

In order to describe the plastic or the viscoplastic material properties of the skeleton material, one has to consider a convenient yield function to bound the elastic domain. Thus, the single-surface yield criterion

$$
\begin{aligned}
F &= \Phi^{1/2} + \beta \mathrm{I} + \epsilon \mathrm{I}^2 - \kappa = 0 \, , \\
\Phi &= \mathrm{II}^D (1 + \gamma \vartheta)^m + \tfrac{1}{2} \alpha \mathrm{I}^2 + \delta^2 \mathrm{I}^4 \, , \\
\vartheta &= \mathrm{III}^D / (\mathrm{II}^D)^{3/2}
\end{aligned}
\tag{44}
$$

for cohesive-frictional materials by Ehlers (1995) is used, cf. Figure 6. Therein, I, II^D and III^D are the first principal invariant of $\mathbf{T}_E^S$ and the (negative) second and third principal invariants of the effective stress deviator $(\mathbf{T}_E^S)^D$. The material parameter sets

$$\mathcal{S}_h = \{\alpha, \beta, \delta, \epsilon, \kappa\}, \quad \mathcal{S}_d = \{\gamma, m\} \tag{45}$$

governs the shape of the yield surface in the hydrostatic plane ($\mathcal{S}_h$) and the deviatoric plane ($\mathcal{S}_d$).

Proceeding either from the viscoplastic approach or from the perfect plasticity concept, the parameters according to (45) are constant during the deformation process and can be computed from standard experimental data by use of an optimization procedure. The above mentioned model can directly be extended to a concept which considers hardening processes, cf. Müllerschön (2000).

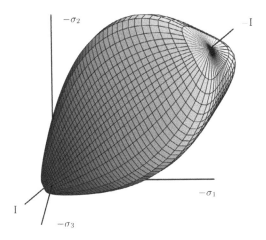

Figure 6: Single-surface yield criterion for cohesive-frictional materials; σ_1, σ_2, σ_3: principal stresses of $\mathbf{T}_E^S$ (tension positive).

Since the associated plasticity concept cannot be applied to frictional materials (Ehlers & Volk 1998), the plastic potential

$$G = \Gamma^{1/2} + \psi_2 I + \epsilon I^2, \quad \Gamma = \psi_1 II^D + \tfrac{1}{2}\alpha I^2 + \delta^2 I^4 \tag{46}$$

(Ehlers & Mahnkopf 1999) is considered, where ψ_1 and ψ_2 serve to relate the dilatation angle to experimental data. From the concept of the plastic potential, it is straight forward to obtain the evolution equation (flow rule) for the plastic strain ε_{Sp} via

$$(\varepsilon_{Sp})_S' = \Lambda \frac{\partial G}{\partial \mathbf{T}_E^S}, \tag{47}$$

where Λ is the plastic multiplier.

In the framework of viscoplasticity using the overstress concept of *Perzyna*-type (Perzyna 1966), the plastic multiplier included in (47) is given by

$$\Lambda = \frac{1}{\eta} \left\langle \frac{F(\mathbf{T}_E^S)}{\sigma_0} \right\rangle^r. \tag{48}$$

Therein, $\langle \cdot \rangle$ are the *Macaulay* brackets, η is the relaxation time, σ_0 the reference stress, and r is the viscoplastic exponent. However, in the framework of elasto-plasticity, where the plastic strains are rate-independent, Λ has to be computed from the *Kuhn-Tucker* conditions

$$F \leq 0, \quad \Lambda \geq 0, \quad \Lambda F = 0 \tag{49}$$

rather than from (48).

3 NUMERICAL TREATMENT

The numerical treatment is based on the weak formulation of the governing field equations together with discretization methods in the space and time domains. Some general reference of the numerical treatment may be found in Helmig (1997), Lewis & Schrefler (1998), Ehlers & Ellsiepen (2001), Ehlers et al. (2001), Ehlers (2002) or Wieners et al. (2002).

Concerning the quasi-static problem under study, it was shown in Section 2.4.1 that the seepage velocities can be eliminated by use of the *Darcy* laws. Thus, $\mathbf{w}_\beta$ loses the status of an independent field variable and the number of equations of the weak formulation reduces to three. In the framework of a standard *Galerkin* procedure (*Bubnov-Galerkin*), the remaining equations are given by the sum of the solid and fluid linear momentum balance equations (mixture momentum balance) multiplied by the test function $\delta \mathbf{u}_S$,

$$\int_\Omega (\mathbf{T}_E^S - p\,\mathbf{I}) \cdot \operatorname{grad} \delta \mathbf{u}_S \, dv = \int_\Omega \rho \mathbf{g} \cdot \delta \mathbf{u}_S \, dv + \int_{\Gamma_t} \bar{\mathbf{t}} \cdot \delta \mathbf{u}_S \, da, \tag{50}$$

by the volume balance relation of the pore-liquid multiplied by the test function δp^{LR},

$$\int_\Omega [(n^L)'_S + n^L \operatorname{div}(\mathbf{u}_S)'_S] \delta p^{LR} \, dv - $$
$$- \int_\Omega n^L \mathbf{w}_L \cdot \operatorname{grad} \delta p^{LR} \, dv = - \int_{\Gamma_v} \bar{v}^L \delta p^{LR} \, da, \tag{51}$$

and finally by the mass balance equation of the pore-gas multiplied by the test function δp^{GR}:

$$\int_\Omega [(n^G \rho^{GR})'_S + n^G \rho^{GR} \operatorname{div}(\mathbf{u}_S)'_S] \delta p^{GR} \, dv - $$
$$- \int_\Omega \rho^{GR} n^G \mathbf{w}_G \cdot \operatorname{grad} \delta p^{GR} \, dv = - \int_{\Gamma_q} \bar{q}^G \delta p^{GR} \, da. \tag{52}$$

Table 1: Full set of parameters.

Parameter	Symbol	Value	Symbol	Value
Lamé constants	μ^S	$5\,583\ \mathrm{kN/m^2}$	λ^S	$8\,375\ \mathrm{kN/m^2}$
Effective densities	ρ^{SR} ρ_0^{GR}	$2\,720\ \mathrm{kg/m^3}$ $1.23\ \mathrm{kg/m^3}$	ρ^{LR}	$1\,000\ \mathrm{kg/m^3}$
Gas phase	$\bar{R}^G$ p_0	$287.17\ \mathrm{J/(kg\,K)}$ $10^5\ \mathrm{N/m^2}$	θ	$283\ \mathrm{K}$
Volume fractions	n_{0S}^S	0.54	n_{0S}^F	0.46
Gravitation	g	$10\ \mathrm{m/s^2}$		
Fluid viscosities and permeability parameters	μ^{LR} π α_{gen} h_{gen} γ_{gen} s_{res}^G	$10^{-3}\ \mathrm{N\,s/m^2}$ 1.0 $2 \cdot 10^{-4}$ 1.5 0.333 0.1	μ^{GR} ξ j_{gen} ϵ_{gen} s_{res}^L	$1.8 \cdot 10^{-5}\ \mathrm{N\,s/m^2}$ 0 2.3 0.5 0.1
Single-surface yield criterion	α γ ϵ m	$1.0740 \cdot 10^{-2}$ 1.555 $4.330 \cdot 10^{-6}\ \mathrm{m^2/kN}$ 0.5935	β δ κ	0.1195 $1.377 \cdot 10^{-4}\ \mathrm{m^2/kN}$ $10.27\ \mathrm{kN/m^2}$
Viscoplasticity	η r	$5 \cdot 10^3\ \mathrm{s}$ 1	σ_0	$10.27\ \mathrm{kN/m^2}$
Plastic potential	ψ_1	1.33	ψ_2	-0.03

Therein, $\bar{\mathbf{t}}$ is the external load vector acting on the *Neumann* boundary Γ_t of the overall medium, $\bar{v}^L = n^L \mathbf{w}_L \cdot \mathbf{n}$ is the efflux of liquid volume through the *Neumann* boundary Γ_v, whereas $\bar{q}^G = \rho^G \mathbf{w}_G \cdot \mathbf{n}$ characterizes the efflux of gaseous mass through the *Neumann* boundary Γ_q; $\mathbf{n}$ is the outward oriented unit surface normal.

The continuous model in its variational form is discretized in the time domain by an implicit *Euler* scheme including a time step control and in the space domain by extended *Taylor-Hood* elements using quadratic shape functions for the solid displacement $\mathbf{u}_S$ and linear shape functions for the effective pressures of the pore-fluids p^{LR} and p^{GR} (Ehlers & Ellsiepen 2001). The space adaptivity is based on a hierarchical refinement and derefinement scheme including as error-indicators the effective solid stresses representing the elastic part of the problem, the plastic part of the strain state representing the accumulated plasticity and the seepage velocities representing the viscosities of the pore-fluids (Ehlers 2002). With respect to the conditions of partially saturated soil, these concept is extended in this paper by the introduction of the effective saturation as additional error indicator.

4 NUMERICAL EXAMPLES

For a validation of the model, the following numerical examples were realized by a computation with the finite element code *PANDAS*. The full set of parameters used for the computations is given in Table 1.

4.1 *Drainage of a soil column*

This problem was considered by Liakopoulos (1964) and therewith, the influence of the pore-gas constituent on the material behaviour of a partially saturated soil can be shown. In this example water is allowed to flow out of a soil column with a height of $h = 1$ m and an intrinsic permeability of $K_{0S}^S = 10^{-13}$ m^2 leading to a liquid *Darcy* permeability of $k_{0S}^L = 10^{-6}$ m/s. The initial conditions for the pore-liquid and the pore-gas saturations are $s^L = 0.9$ and $s^G = 0.1$. From this state, the water table decreases to the bottom of the column due to gravity effects.

The upper part of Figure 7 shows the pore-liquid saturation s^L versus time for the computation under the assumption of a static gas-phase (*Reynold*'s assumption: $p^{GR} \equiv 0$) and the lower part

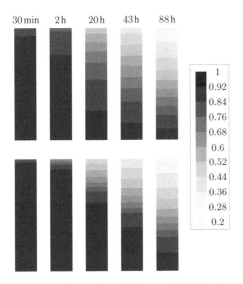

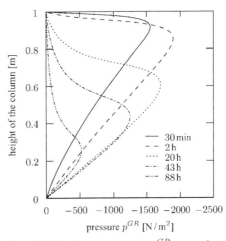

Figure 7: Saturation of the pore-liquid during the drying process, upper series: biphasic model ($p^{GR} = 0$), lower series: triphasic model.

Figure 8: Effective pore-pressure p^{GR} versus time.

257

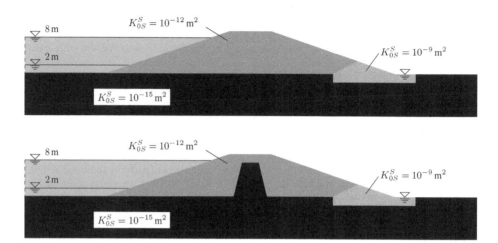

Figure 9: Embankment with and without a seal unit.

for the triphasic formulation. It can be realized that the resulting negative effective pressure of the pore-gas, cf. Figure 8, affects the fluid-flow situation significantly. The maximum is around $p^{GR} = -2000\,\mathrm{N/m^2}$, but it disappears according to the gas permeability of the given material.

4.2 *Flow through an embankment*

As a further example, the flow through an embankment is considered. The height of the embankment is about 10 m and the slope gradient on both sides about $s = 1/3$, cf. Figure 9. The "impermeable" ground-level is governed by an intrinsic permeability of $K_{0S}^S = 10^{-15}\,\mathrm{m^2}$ leading to a liquid *Darcy* permeability of $k_{0S}^L = 10^{-8}\,\mathrm{m/s}$. The embankment has an intrinsic permeability of $K_{0S}^S = 10^{-12}\,\mathrm{m^2}$ ($k_{0S}^L = 10^{-5}\,\mathrm{m/s}$), whereas the filter at the right-hand side of the embankment has a permeability of $K_{0S}^S = 10^{-9}\,\mathrm{m^2}$ ($k_{0S}^L = 10^{-2}\,\mathrm{m/s}$). To determine the influence of a seal unit on the flow behaviour of the pore-fluids, the embankment is also considered with a seal unit. In this

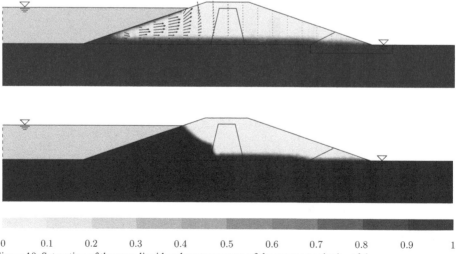

Figure 10: Saturation of the pore-liquid and vector arrows of the seepage velocity of the pore-gas.

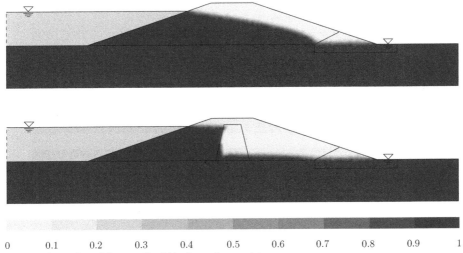

Figure 11: Saturation of the pore-liquid in the stationary state.

case, the seal unit has the same intrinsic permeability as the impermeable ground level. From the initial stationary state with a water level of 2 m over the impermeable ground-level on the left-hand side of the embankment, the water level is rapidly increased to 8 m, cf. Figure 9.

Figure 10 shows the pore-fluid flow situation at different times. The upper part shows the saturation of the pore-liquid, where the vector arrows indicate the seepage velocity of the pore-gas three hours after the water level has been increased. It can be recognized that the pore-liquid is displacing the pore-gas. Five days later, the water front is rising up to the seal unit, cf. lower part of Figure 10.

In Figure 11, the pore-liquid saturation in the final stationary state is shown. It can be realized that the seal unit prevents the increase of the free water level in the right part of the embankment and so, the air side of an embankment with a seal unit is securer against destruction as the air side of the same embankment without a seal unit. The final finite element mesh of the computation without a seal unit is given in Figure 12. The elements are refined in that zones where high gradients of the effective saturation occur.

If the embankment is constructed without a filter or a seal unit, the pore-liquid will flow out of the embankment over the slope on the air side. The result is an instability of this slope due to the effect that the soil beneath the water-table is under buoyancy and this force can not be balanced on the slope by the self-weight of the soil. The accumulated plastic strains start to localize and to form a shear band leading to a destructing of the embankment, cf. Figure 13. This situation can be prevented by an extra load on the slope resulting, e. g., from sandbags.

Another situation leading to a destruction of an embankment is a rapidly decreasing water level. In this situation, the accumulated plastic strains start to localize and to form a shear band on the water side, cf. Figure 14.

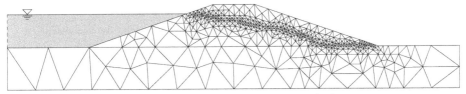

Figure 12: Final mesh in the stationary state (embankment without a seal unit).

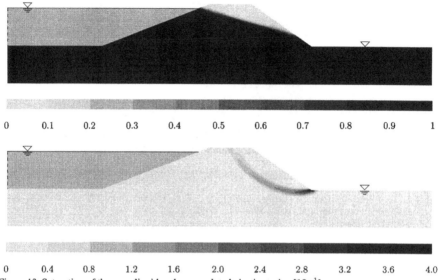

| 0 | 0.1 | 0.2 | 0.3 | 0.4 | 0.5 | 0.6 | 0.7 | 0.8 | 0.9 | 1 |

| 0 | 0.4 | 0.8 | 1.2 | 1.6 | 2.0 | 2.4 | 2.8 | 3.2 | 3.6 | 4.0 |

Figure 13: Saturation of the pore-liquid and accumulated plastic strains $[10^{-1}]$.

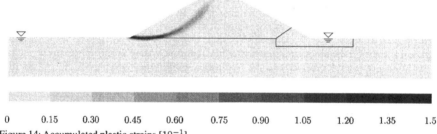

| 0 | 0.15 | 0.30 | 0.45 | 0.60 | 0.75 | 0.90 | 1.05 | 1.20 | 1.35 | 1.5 |

Figure 14: Accumulated plastic strains $[10^{-1}]$.

5 CONCLUSION

In the present contribution, a triphasic model for partially saturated soil was developed on the basis of the *Theory of Porous Media*. Furthermore, a thermodynamically consistent framework to describe the deformation of the solid skeleton and the motion of the pore-fluids relative to the solid skeleton was presented. The model consists of a materially incompressible solid skeleton, a materially incompressible pore-liquid and a materially compressible pore-gas. For a description of the interactions between the constituents, the capillary-pressure-saturation relation, the relative-permeabilities-saturation relation following van Genuchten (1980) and constitutive equations for the momentum supply terms leading to the *Darcy* filter law were used. Taking into account an elasto-plastic or an elasto-viscoplastic material behaviour of the skeleton material, a realistic model of partially saturated soil is obtained.

The model was treated numerically within the framework of the Finite Element Method. For a discretization in time and space, the implicit *Euler* scheme and extended *Taylor-Hood* elements were used.

By a computation of a drying soil column with and without a static gas-phase, it could be shown that the influence of the pore-gas on the material behaviour of a partially saturated soil is not generally negligible, i.e. the assumption of a static gas-phase is restricted to particular initial-boundary-value problems. Furthermore, the flow through an embankment was simulated numerically. Thereby, it could be shown that the model is capable to describe the realistic pore-fluids flow situation and the interaction between the pore-fluids and the solid skeleton deformation and localization.

It is planned to extend the model by mass exchanges between the constituents and hysteretic capillary-pressure-saturation relations and to validate the model by experiments in cooperation with the DFG research group *Mechanics of Partially Saturated Soils* (www.iws.uni-stuttgart.de/Hydrosys/Forschung/dfg_forschergruppe).

REFERENCES

Bishop, A. W. 1959. The principle of effective stress. *Teknisk Ukeblad* 39: 859–863.

Bowen, R. M. 1976. Theory of mixtures. In A. C. Eringen (ed.), *Continuum Physics, Volume III*: 1–127. New York: Academic Press.

Bowen, R. M. 1980. Incompressible porous media models by use of the theory of mixtures. *Int. J. Engng. Sci.* 18: 1129–1148.

Brooks, R. N. & Corey, A. T. 1966. Properties of porous media affecting fluid flow. *J. Irrig. Drain. Div. Am. Soc. Civ. Eng.* 92: 61–68.

Ehlers, W. 1993. Constitutive equations for granular materials in geomechanical context. In K. Hutter (ed.), *Continuum Mechanics in Environmental Sciences and Geophysics, CISM Courses and Lectures 337*: 313–402. Wien: Springer-Verlag.

Ehlers, W. 1995. A single-surface yield function for geomaterials. *Arch. Appl. Mech.* 65: 246–259.

Ehlers, W. 2002. Foundations of multiphasic and porous materials. In W. Ehlers & J. Bluhm (eds.), *Porous Media: Theory, Experiments and Numerical Applications*: 3–86. Berlin: Springer-Verlag.

Ehlers, W. & Ellsiepen, P. 1998. PANDAS: Ein FE-System zur Simulation von Sonderproblemen der Bodenmechanik. In P. Wriggers, U. Meißner, E. Stein & W. Wunderlich (eds.), *Finite Elemente in der Baupraxis: Modellierung, Berechnung und Konstruktion*: 391–400. Berlin: Ernst & Sohn.

Ehlers, W. & Ellsiepen, P. 2001. Theoretical and numerical methods in environmental continuum mechanics based on the Theory of Porous Media. In B. A. Schrefler (ed.), *Environmental Geomechanics, CISM Courses and Lectures 417*: 1–81. Wien: Springer-Verlag.

Ehlers, W., Ellsiepen, P. & Ammann, M. 2001. Time- and space-adaptive methods applied to localization phenomena in empty and saturated micropolar and standard porous materials. *Int. J. Numer. Methods in Engng.* 52: 503–526.

Ehlers, W., Ellsiepen, P., Blome, P., Mahnkopf, D. & Markert, B. 1999. *Theoretische und Numerische Studien zur Lösung von Rand- und Anfangswertproblemen in der Theorie poröser Medien, Abschlußbericht zum DFG-Forschungsvorhaben Eh 107/6-2*. Bericht Nr. 99-II-1 aus dem Institut für Mechanik (Bauwesen), Universität Stuttgart.

Ehlers, W. & Mahnkopf, D. 1999. Elastoplastizität und Lokalisierung poröser Medien bei finiten Deformationen. *ZAMM* 79: S543–S544.

Ehlers, W. & Volk, W. 1998. On theoretical and numerical methods in the Theory of Porous Media based on polar and non-polar elasto-plastic materials. *Int. J. Solids Structures* 35: 4597–4617.

Eipper, G. 1998. *Theorie und Numerik finiter elastischer Deformationen in fluidgesättigten porösen Festkörpern*. Dissertation, Bericht Nr. II-1 aus dem Institut für Mechanik (Bauwesen), Universität Stuttgart.

Finsterle, S. 1993. *Inverse Modellierung zur Bestimmung hydrogeologischer Parameter eines Zweiphasensystems*. Technischer Bericht der Versuchsanstalt für Wasserbau, Hydrologie und Graziologie der Eidgenössischen Technischen Hochschule Zürich.

Helmig, R. 1997. *Multiphase Flow and Transport Processes in the Subsurface: A Contribution to the Modeling of Hydrosystems*. Berlin: Springer-Verlag.

Lewis, R. W. & Schrefler, B. A. 1998. *The Finite Element Method in the Static and Dynamic Deformation and Consolidation of Porous Media* (second ed.). Chichester: Wiley.

Liakopoulos, A. C. 1964. *Transient Flow Through Unsaturated Porous Media*, Ph. D. thesis. University of California, Berkeley.

Müllerschön, H. 2000. *Spannungs- Verformungsverhalten granularer Materialien am Beispiel von Berliner Sand*. Dissertation, Bericht Nr. II-6 aus dem Institut für Mechanik (Bauwesen), Universität Stuttgart.

Perzyna, P. 1966. Fundamental problems in viscoplasticity. *Adv. Appl. Mech. Eng.* 9: 243–377.

Truesdell, C. 1984. Thermodynamics of diffusion. In C. Truesdell (Ed.), *Rational Thermodynamics* (second ed.): 219–236. New York: Springer-Verlag.

Truesdell, C. & Toupin, R. A. 1960. The classical field theories. In S. Flügge (Ed.), *Handbuch der Physik, Volume III/1*: 226–902. Berlin: Springer-Verlag.

van Genuchten, M. T. 1980. A closed-form equation for predicting the hydraulic conductivity of unsaturated soils. *Soil Science Society of America Journal* 44: 892–898.

Vogler, M. 1999. *Einfluß der Kapillarität auf die Mehrphasenströmung bei der Sanierung von Mineralölschadensfällen im Boden*. Dissertation, Mitteilungen des Institutes und der Versuchsanstalt für Geotechnik, Heft 45, Technische Universität Darmstadt.

Wieners, C., Ammann, M., Diebels, S. & Ehlers, W. 2002. Parallel 3-d simulations for porous media models in soil mechanics. *Computational Mechanics* 29: 75–87.

Author index

T - #0281 - 101024 - C0 - 246/174/15 [17] - CB - 9789058095633 - Gloss Lamination